WORLDS WITHOUT END

The Historic Search for Extraterrestrial Life

First published 1999

PUBLISHED IN THE UNITED KINGDOM BY:

Tempus Publishing Ltd
The Mill, Brimscombe Port
Stroud, Gloucestershire GL5 2QG

PUBLISHED IN THE UNITED STATES OF AMERICA BY:

Tempus Publishing Inc.
2A Cumberland Street
Charleston, SC 29401

Tempus books are available in France, Germany and Belgium
from the following addresses:

Tempus Publishing Group	Tempus Publishing Group	Tempus Publishing Group
21 Avenue de la République	Gustav-Adolf-Straße 3	Place de L'Alma 4/5
37300 Joué-lès-Tours	99084 Erfurt	1200 Brussels
FRANCE	GERMANY	BELGIUM

British Library Cataloguing in Publication Data.
A catalogue record for this book is available from the British Library.

ISBN 0 7524 1450 X

Typesetting and origination by Tempus Publishing.
PRINTED AND BOUND IN GREAT BRITAIN.

Contents

Introduction

There are many books about the possibility of worlds other than ours, and of life, intelligent or otherwise, inhabiting them. Nevertheless, there are very few histories of this idea, although people have speculated about it for 2500 years. Chronicles have now begun to appear; any writer on the subject will be in debt for many years to come to the scholarship of Michael Crowe, Steven Dick and Karl Guthke, details of whose works appear in the bibliography and elsewhere. None of these books, remarkable for the density of their erudition, encompasses the entire period in question; that is what this modest contribution to the discourse attempts to accomplish.

The title of this book is not original. Considering the vastness of the subject, the possibility of worlds elsewhere in the Universe, the verbal permutations of straightforward titles are curiously few; hence the many books asking 'are we alone?', 'is anyone out there?' and so on. The title *Worlds Without End* was employed by N. J. Berrill (1964) and earlier by Sir Harold Spencer Jones, Astronomer Royal (1935-55), an illustrious scientist who kept the idea of pluralism alive at a time when it was generally unfashionable; the shared title has therefore a sound pedigree.

Usages

The dates of the various authorities cited can be found in the Index; here and there they have been put in the main text also, usually where a chronological landmark is necessary. Biblical quotations are from the King James' Authorised Version of the Bible, the text most commonly employed in the historical debates on a plurality of worlds.

In keeping with astronomical usage, the capitalised Sun, Moon, Galaxy and Universe refer to those with which we are familiar, our Sun, our Galaxy, etc. Other suns, moons and galaxies are not so typographically distinguished.

Many technical terms are defined in context; they and others will also be found in the Glossary.

Quotations have occasionally had their spelling modernised, otherwise they are left generally as written.

Illustration numbers are in **bold** type; footnote numbers refer to material placed at the end of each chapter.

Front cover: *Has Man a Brother in the Skies?* From Harmsworth's *Popular Science*, edited by Arthur Mee (1912).

Acknowledgements

In addition to drawing on the invaluable spadework of the pioneers mentioned above I have consulted many of the works cited in the Bodleian Library at Oxford, and express gratitude for the swift and courteous service of its staff. I must also thank the following for invaluable assistance: the British Astronomical Association; the BBC; the National Astronautics and Space Administration, of the USA; the Royal Astronomical Society; the Royal Astronomical Society of Canada; the Royal Greenwich Observatory; the Royal Swedish Academy of Sciences; the information and publicity departments of Woking Borough Council.

Most illustrations are from private collections; others have their sources recorded in the appropriate place.

I must also thank my wife, Jenny, for her impeccable editing, and for raising awkward and therefore fruitful questions. This book is dedicated to her.

1 A plurality of worlds, from Plato to NATO

The concept of other, inhabited worlds, of a 'plurality of worlds' is an old one. The first documented pluralist source extant dates at least from the pre-Socratic philosophers of ancient Greece. The story of pluralism is, however, fragmented; as an intellectual fashion it has waxed and waned.

Pluralism has involved a wide range of people and cultural forms: Epicurus, Aquinas, Descartes, Kant, H.G.Wells, *Star Trek*, the US space program, the Holy Inquisition, the Swedenborgian Church, and contemporary cults surrounding Unidentified Flying Objects and abductions by aliens.

In the last two centuries, interest in pluralism has grown steadily in sophistication and popularity. If opinion polls are reliable guides, perhaps the majority of people in the modern industrialised world believe in the possibility, or actuality, of extraterrestrial beings. It has been suggested that pluralism is the 'myth of the modern age'[1], permeating science and popular culture. One of the leading historians of the field, Steven J. Dick, suggested that pluralism is, in effect, the axis around which modern science moves, its assumptions and enquiries suffusing astronomy, cosmology, the life sciences and space science.[2]

Many cultures have attributed other-worldly qualities to the constellations or individual stars of the night sky (**1**). To the Chumash of California 'sky beings were actually people who had ascended from Earth long ago to escape death in a primeval flood'.[3] The Pitdjandjara of Western Australia held, as did Greek mythology, that the Pleiades were 'originally a group of women who lived on Earth'.[4]

A more secular concept developed also in classical Greece — that the Earth was part of a *kosmos*, a cosmic system of celestial machinery. During the period of Greek civilisation, 600BC – AD200 this concept changed and evolved. The notion of *aperoi kosmoi*, an infinity of worlds, emerged as a minority view. The usual view, held for example by Plato and Aristotle, was that there was one world-system centred on the Earth around which the Sun, Moon, planets and stars revolved, for its pleasure and convenience.

Changing usages tell their own story. From the Greeks to the late Middle Ages 'world' meant the whole cosmic system centred on and including our Earth (**2**). Only later did 'world' come to mean a discrete earth, such as ours. After *c*1600 people spoke of a 'plurality of worlds' implying 'many other worlds like our own planet'. In modern times an interest

1   *Pluralist fantasy: the stars as glittering jewels of celestial super-beings from* Un Autre Monde *by Grandville (J-I-I Gérard), 1844.*

grew in the life-forms, if any, that dwelt on these worlds; Victorians used the phrase 'life on other worlds'. More recently we have focused on one feature of this life, the search for 'extraterrestrial intelligence'.

Some enduring features of pluralism had been established by the time classical civilisation disintegrated. First, for most of its history, pluralism was to be sectarian, the persuasion of a small number of creative intellectuals. Secondly, pluralism has generally been a source of controversy although gradually, from the Renaissance, it became more popular and widespread until it achieved majority status. But it still has its critics and opponents. Thirdly, it was (and remains) a house containing many mansions: it speaks of many worlds; an infinity of worlds; worlds succeeding each other in time, or in space; extraterrestrial beings of many forms and qualities. It has been discussed not only in science and philosophy, but also in literature, drama and music from the time of Plutarch to modern space operas.

Early pluralism had another characteristic which has lasted: it was a largely Western phenomenon. The contemporary global interest in 'life on other worlds' is the direct descendant of the pioneer theorising of Greek philosophers, spreading with Western science and technology. It endured through the Middle Ages, the Renaissance and the Enlightenment, changing and adapting on its long course, until it was eventually built into the popular culture of the industrialised world, particularly the culture of its pre-eminent location, the United States. There were other pluralisms, in India and China for example. But, unlike Greek pluralism, these were not situated within cosmologies or frameworks of scientific and philosophical enquiry that developed with the density, elaboration and scale of the Hellenic version, and they did not contribute much to the development of modern pluralism. Western pluralism, speculation about life in infinite space, may also

2　*The medieval cosmos, with the Earth at the centre and the habitation of the Elect lying outside the* primum mobile *which kept the system turning; from Peter Apianus'* Cosmographicus Liber, 1524.

have been another manifestation of that restless, Faustian spirit that urged on other characteristic occidental activities: exploration, mountaineering, imperialism, space travel and other sources of excitement and grief.

Pluralism remained for many centuries more metaphysical than scientific in its tone and emphasis. Until the Renaissance it depended on speculation and guesswork; only later — and very modestly at first — did scientific investigation provide it with a firm basis. The theory of Copernicus, that the Earth was not the centre of the cosmos, and the invention of the telescope in the early seventeenth century were the first notable causes of scientific pluralism. When, for example, people observed mountains on the Moon (3), or four (later many more) moons orbiting Jupiter they could see for themselves that Earth had no special monopoly of these features. Even then pluralists continued to argue their case from analogy and probability: if the Moon had mountains then surely it was a world like the Earth? Surely amongst the millions of stars that glittered in the spyglass there must be some other worlds? Scientific pluralism and metaphysical pluralism have continued to exist side-by-side ever since, but from the mid-nineteenth century scientific pluralism has been in the ascendant, greatly strengthened by developments in astronomy and the life sciences.

Some features of later pluralism were not part of the Greek bequest. Although pluralism was a minority philosophy in Greece and had to contend with a prevailing anti-pluralist cosmology and the complexities of pagan religion, it did not have to face the formidable intellectual and secular power of the Christian Church. During and immediately after the Middle Ages pluralism was to have a difficult relationship with Christianity, although it benefited greatly from this struggle which sharpened thinking and kept pluralists on their toes.

3   The lunar crater of Eratosthenes from Half Hours in the Air and Sky *(1877) — Galileo's discovery of supposedly Earth-like features on the surface of the Moon greatly encouraged pluralist thought.*

Christian theology traditionally raised two objections to pluralism: first that this world, and humanity, are special creations of God and have not been replicated; secondly that God made Himself incarnate on the Earth. The existence of other inhabited worlds therefore raised difficult questions about the uniqueness of the Incarnation, a problem with which Islam, Judaism or Hinduism have not had to grapple.[5] These religions have generally been at ease with the concept of a plurality of worlds but their homelands did not develop it as far as Christendom, where a harsher intellectual climate seems to have stimulated its growth. Pluralism eventually became a prop to the faith of many Christians to whom a plurality of worlds in a vast Universe demonstrated the omnipotence of a God who would not create heavenly bodies in vain but would populate them as He had the Earth. In our more secular age the 'moral economy' of other worlds has ceased to be a central issue in pluralist debate as it was from the Renaissance to mid-Victorian times; this loss has impoverished pluralist discourse, leaving it incomplete and vulnerable.

Nor did the Greeks develop a popular, let alone a mass pluralism. After the Renaissance the intelligentsia — astronomers, philosophers, theologians, poets, and novelists — took an increasing interest in pluralism. As literacy spread so did popular pluralist essays and books; by the 1730s pluralism had become an intellectual fashion in the polite society of Western Europe. Mass pluralism first appeared in the 'Great Moon Hoax' of 1836, the original, successful 'media hype' based on misinformation. It reappeared later in the nineteenth century as education spread, embodied then as now in popular science and science fiction.

Pluralism's later, strong, links with science, particularly astronomy, have not been without risk. Whereas metaphysical pluralism was nourished by the copious springs of philosophical ingenuity, scientific pluralism has had to respect the canons of hard science. If the evidence went against pluralism, so much the worse for a plurality of worlds. At the end of the nineteenth century the alliance with the natural sciences seemed to bear fruit: it appeared that a great network of artificial canals had been observed on Mars. But when this theory was demolished by science, pluralism took a hard blow. It went quiet between the wars, 1919-39, its main refuge becoming science fiction, which may have have further weakened its case in the eyes of orthodox science.

Since the 1950s, pluralism as a cultural form has grown massively to attain the level and status of a 'defining myth'. The underlying causes of its growth have both multiplied and intensified, particularly its links with science. There has been an astonishing renaissance of astronomy and cosmology, and a similar growth in interest in the life sciences.

Pluralism is relevant to many concerns of contemporary science and has in turn generated its own formal, organised, bureaucracy; a world of international conferences, academic papers, protocols and manoeuvring for research grants. Even so, the mystery at the heart of pluralism seems to remain intact and to drive activity forwards.

Education and relative affluence have also had their own effects. The numbers of amateur astronomers have multiplied, but the most noticeable attraction of pluralism has been apparent in the generation of science fiction and its latter-day offspring: space operas, videos, fan networks; and of people who follow UFO news and wonder if our Earth was once visited by 'ancient astronauts'. This popular interest has also influenced modern art, sculpture (4), music, even gardening and home decorating.

Why pluralism, in any case? Why this time-hallowed interest in the possibility of life on other worlds than ours? Perhaps because it offers some consolation or escape from the crass materialism, consumerism and commercial manipulations of contemporary society, particularly for people who value the qualities of wonder and imagination which have coloured the subject over the ages. But more probably because it necessarily touches, as it always has done, on the big questions: Are we unique? Why are we here? How should we behave towards other people, or other creatures of Nature? And how are all things made for man?

References
1   Karl S. Guthke, *Der Mythos der Neuzeit*, 1983; trans *The Last Frontier*, 1990.
2   Steven J. Dick, *The Biological Universe*, 1996.
3   E. Hadingham, *Early Man and the Cosmos*, 1983.
4   Christopher Walker, ed, *Astronomy Before the Telescope*, 1996.
5   Nor, for related reasons Socinians, Arians and Unitarians; the Unitarian contribution to pluralist debate has, however, not been extensive.

4   *Pluralism in public art: Michael Condron's steel sculpture 'The Martian' (1998) commissioned by Woking Borough Council to commemorate the centenary of H. G. Wells'* The War of the Worlds, *written in that Surrey town.*

2 Greece and the East

Classical Greece

Anaximander (sixth century BC) was the first eminent pluralist. His pluralism arose from an interest in astronomy. He learned from his mentor, Thales of Miletus, that the constellation we now refer to as Ursa Major (also known as The Great Bear, Charles' Wain, the Dipper) sank partly over the horizon in Egypt, which it did not in his native Miletus, and concluded with a lateral thought of great potential that the Earth was cylindrical, set within an enveloping spherical sky. Rudolf Thiel[1] thought that Anaximander's deduction was 'one of the most momentous conclusions in all antiquity'. Of the pluralism of Anaximander and the other early 'pre-Socratic' philosophers we know little beyond their basic interest and a few assertions.[2] Thales, himself an accomplished mathematical astronomer, thought that the Moon was constructed similarly to the Earth. Anaximander suggested that there might be an infinity of worlds set apart from each other[3] an idea shared by his pupil Anaximenes who adopted a less sophisticated cosmology than that of his teacher — a cosmos with a flat, disc-like Earth overarched by a vault to which the stars were fixed. But beyond this vault might lie others, a pluralist elaboration. For all their sophistication these ideas represented the taste and achievement of a minority; they had to coexist with a widespread, traditional mythology.

Nevertheless, the Greek notion of *aperoi kosmoi*, an 'infinity of worlds', from which modern pluralism is ultimately descended was launched in Ionia in the fifth century BC. In its original and simple form it was taken up or repeated by other Greek thinkers such as Archelaus, preceptor of Socrates, and his disciple Xenophanes who stated that the Moon, which he believed to be eighteen times the size of the Earth, was inhabited. Although the cosmologies of these and other Greek philosophers were often complex, they all rested on a single, shared perception that the 'world' was geocentric, an Earth-centred system of heavenly bodies. Few Greeks escaped from this mind-set: Aristarchus of Samos (*c*320-250 BC) did, but was accused of 'impiety' for daring to suggest that the Earth turned on its axis and orbited the Sun,[4] and that the Earth did not lie at the centre of the cosmos.

An alternative way of perceiving an infinity of *kosmoi* was to see them as succeeding each other in time, rather than stretching away infinitely into space. This was suggested by Heraclitus of Ephesus who thought the cosmos and all within it was in a constant state of flux or change; it renewed itself in an infinite succession of 10,800-year cycles. Later, the Stoic school of philosophy developed a related idea, the periodic renewal or rejuvenation of the cosmic system by massive conflagrations. The Stoic belief that the heavenly bodies, the Sun, Moon and stars were spherical like the Earth, provided the basis of pluralism by analogy, a long-enduring characteristic of the subject.[5]

Pythagoras and his school (sixth – fourth centuries BC) also contributed to the development of pluralism; they varied the strictly geocentric line of Greek cosmology by positing, at least in their earlier phase, a 'central fire', around which the Earth, Sun, Moon, planets and stars revolved. Between Earth and the central fire lay *antichthon*, the counter-Earth, about the habitability of which they remained ambiguous. One of the later compilations of Greek philosophy, the *Placita* of Pseudo-Plutarch, records, however, the Pythagorean belief that 'the moon... is inhabited like the Earth and contains animals of a larger size and plants of a rarer beauty than our globe affords'.

Another Pythagorean, Petron, suggested with unusual precision for an early pluralist that there were 183 worlds arranged in a triangle.[6] The chief concern of the Pythagoreans was nevertheless to establish mathematical relationships within the cosmos, even characterising the cosmos. To this end they proposed that each of the planets in its passage around the central fire emitted a sound, the 'music of the spheres', an appealing concept which was occasionally to be taken up by pluralists. According to Alexander of Aphrodisias (third century BC):

> The Pythagoreans said that the bodies of the planetary system revolve around the centre at distances related to mathematical proportions... the slower ones emit deeper sounds as they move, the quicker ones higher sounds. These sounds depend on the ratios of the distances, which are so proportioned that the combined effect is harmonious....

Pythagorean suggestions of an inhabited Moon were matched by others; in the *Orphics*, the poetry attributed to Orpheus, we read of 'Mountains, men and cities in the Moon.' Similarly, the pluralist analogy of some Pythagoreans that each planet resembled the Earth and had its own atmosphere and Earthly trappings, was attributed also to the astronomer Heraclides who spoke of an infinite cosmos centred on a revolving Earth.

These fragments demonstrate that Greek ideas of a possible infinity, or at least a plurality of worlds were more elaborate and advanced than that of any other society prior to the European Renaissance. Even so, for all their originality the Greeks — and particularly the earlier, pre-Socratic Greeks, did not offer a coherent, developed pluralist theory. The nearest classical Greece came to a theoretical system that might underpin and embrace pluralism was *atomism*, derived from the thinking of Leucippus (mid-fifth century BC) and his pupil Democritus. Their doctrine was that: the material Universe was composed of indivisible atoms; the atomist cosmos was evolutionary and automatic; it required no Divine spark to set it going or to sustain it on its course. Also, it was extremely hospitable to concepts of a plurality of worlds. The eminent atomist Epicurus (341-270BC) put the matter thus in his 'letter to Herodotus':

> There are infinite worlds both like and unlike this world of ours... atoms being infinite in number... are borne far out into space... there nowhere exists an

obstacle to the infinite number of worlds... in all worlds there are living creatures and plants... we see in this world.

This, amongst the clearest statements of Greek pluralism, echoed other atomist propositions. Leucippus and Democritus were themselves said to be pluralists, although we have no record of pluralist theory attributable to them. We learn, however, from Diogenes Laertius writing much later (third century AD) that 'Leucippus holds that the whole is infinite... hence arise innumerable worlds', and from Hippolytus, a theologian, that Democritus thought that there might well be other worlds; some might lack a sun or moon, others might have suns or moons larger than ours, and yet others bear no living creatures, plants or moisture — a fair description of the surface of Mars as seen by the *Sojourner* vehicle in late 1997.

There were subtle differences within atomist pluralism. To Democritus a plurality of worlds was the outcome of an infinite cosmos containing an infinity of atoms; to Epicurus the plurality of worlds grew from a surplus of atoms 'not used up on one world or a limited number of worlds'. He was, nevertheless in agreement with Democritus in his belief that these other worlds contained living creatures and plants 'and other things with which we are familiar in this world...'. This plurality, like that of Anaximander or Xenophanes, was of entire world systems set apart from our own, Earth-centred one. Their asserted existence grew from atomist logic, not from direct observation. Indeed, although these worlds existed to the satisfaction of the atomist mind, it was understood that they could never be seen or experienced by the inhabitants of Earth.

Atomism remained the most complete and internally logical theory regarding the plurality of worlds to emerge from classical Greece. It had some noteworthy refinements, however. The atomist Metrodorus of Chios, disciple of Epicurus, put his pluralist thoughts into metaphorical form:

> It would be strange if a single ear of corn grew in a large plain or there were only one world in the infinite. And that worlds are infinite follows from the causes being infinite.

In so saying, Metrodorus was amongst the first to launch that which became known in due course as the *principle of plenitude*, the idea that any, even all, possibilities can and will be fulfilled at some point and time, a principle clearly supportive of pluralism and of those who were to argue for a plurality of worlds on grounds of *probability*.

Metrodorus also pioneered the probabilistic metaphor much favoured by later pluralists. The Russian pioneer of space science, himself a moderate pluralist, Konstantin Tsiolkovsky wrote in 1895: 'Is it conceivable for one apple tree in the infinite orchard of the Universe to bear fruit whilst innumerable other trees have nothing but foliage?' — a clear echo of Metrodorus.

As Hellenic culture was absorbed into that of the Roman Empire, pluralism was carried along with it; it had its Roman adherents in due course. The most prestigious of these, the poet Lucretius (*c*99–55 BC), developed the philosophy of Epicurus:

> Empty space extends without limit in every direction... it is in the highest degree
> unlikely that this Earth and sky is the only one to have been created... in other
> regions there are other earths and various tribes of men and breeds of beasts.

The work in which this appeared, *De Rerum Natura*, was to have great influence in the
Renaissance some 1500 years later by being one of the chief means by which Greek
atomism came to colour European science and also to assist the relaunch of pluralism after
its slumbers in the Middle Ages. Other Romans showed a sympathetic interest in
pluralism; Plutarch (*c*AD46-120), for example thought that there might be life on the
Moon.

Lucian of Samosata, procurator of part of Egypt, probably influenced by his sympathy
to atomism, described two imaginary space-trips. These were historically significant as the
earliest science fiction; the first known employment of pluralist theory for fictional
purposes, a literary form which has endured into modern times. In *Icaro-Menippus*, the
traveller Menippus acquires wings and flies to Olympus via the Moon. In *True History*,
Lucian himself visits the Moon with fifty armed companions, well-honed Greek athletes,
and becomes involved in a space-war between the inhabitants of the Moon and the Sun
over the colonisation of 'the morning star', Venus. He then returns to Earth via the Sun
for further adventures.

Plutarch also wrote a pioneering work of science fantasy, involving the moon: *Facies in
Orbe Lunare* 'Of the face which appears on the orb of the moon', is framed around an
exchange of views between eight friends, one of whom poses the question: 'Is the Moon
inhabited?' The answer involves a discussion about the hostile conditions facing life on
the Moon including 'twelve summers each year' and a speculation that the souls of dead
people inhabit our satellite — the first mention of a theme which was to recur in pluralism
and science fiction in modern times.[7]

Whereas the works of Greek pluralists available to contemporary readers are scarce,
those of their critics are more numerous, complete and prestigious. Two eminent sources
of anti-pluralism still have many of their works in print: Plato (428–348 BC) and Aristotle.
Plato's opposition to pluralism arose, as did that of Aristotle, as a logical outcome of his
cosmology(5). To Plato, in his *Timaeus*, the cosmos was the single creation of a single
creator, therefore 'There is and ever will be one only-begotten and created heaven.'

Aristotle's opposition to pluralism, expressed in *De Caelo* 'On the Heavens' was more
complex. Like many Greek philosophers he was closely engaged with the question of
what, ultimately, constituted the Universe. Whereas the atomists imagined that it was
made up of indivisible atoms, Aristotle and others thought that it was composed of four
fundamental elements: earth, air, fire and water. All entities were in turn composed of
these substances, or combinations of them. Here the anti-pluralist argument became
convoluted, but effective in its own terms. Fire, for example, had a natural tendency to rise
up and away from the Earth (in this sense meaning our planet, the centre of the cosmos).
A lump of earth (ie, soil) will tend to fall down towards it. But if there were another world
somewhere, what then? Would the lump fall towards the centre of *that* world? If so, it
would implicitly and explicitly deny Earth as being the centre of the cosmos. If not, then
arrangements on the 'other world' would be topsy-turvy and unworkable. The conclusion

therefore must be that a given lump of earth, or leaping flame, could not exist in any system composed of more than one world.

Aristotle had other anti-pluralist weapons. In the generally accepted Greek cosmological model, the Sun, Moon, planets and stars circled around Earth, fixed to spheres. Outside these body-bearing spheres there turned the *primum mobile*, a kind of cosmic flywheel or sphere which imparted motion to the rest of the system. In his *Metaphysics*, Aristotle noted that more than one world would lead to the absurdity of there being a plurality of prime movers, unacceptable in logic and offensive to any religious faith that posited one cosmos and one Creator. Similarly, Aristotle held that this world — in the sense 'world-system', contained all matter in existence, that is all 'particulars'. A plurality of worlds would suggest a possible infinity of particulars rather than the complete, finite, number of particulars which he held to be the case.

The standard Greek geocentric cosmology adopted by Aristotle had much to recommend it in a rough and ready way. The Sun appears, after all, to rise in the East and set in the West. Our language still embodies this perception, we speak of 'sunrise', not 'Earthdip'. In its basic form it could not explain certain astronomical oddities such as the differential speeds of the planets, and the retrograde movements of Mars. But, if one were prepared to live with some tweaks and convolutions, this cosmology, especially as it was summarised in the amply-detailed *Almagest* of Ptolemy (first century AD) would serve both common sense and everyday usage for some 1500 years. No material advances in pluralism were to be made from the early Roman empire until the European Renaissance; its few adherents had their work cut out squaring pluralism, and its implied critiques of Ptolemaic cosmology and Aristotelian science, with the tenets of the Christian Church. In that God's close interest in this Earth and humanity was fundamental to Christian faith, geocentric notions well suited its theology and assumptions; speculation about extraterrestrial life did not.

The Orient

Although Greek pluralism had been intellectually fertile and subtle, the tradition it launched was not to be the only one. The other ancient pluralisms were not, however, linked with cosmologies ultimately capable of extensive development and verification. The ability of the Greek tradition to do this has brought it, or rather its distant descendants, to a culturally dominant position.

At least one school of Buddhism was amongst the sources of Oriental atomism which was not unlike its Greek counterpart in important respects, although none of the Eastern atomist traditions developed Greek-type pluralism.[8] Buddhism was favourably aligned towards pluralism, although not being inclined to dwell heavily on worldly matters, it did not develop a scientific basis for it.

The concepts of infinity and of a plurality of worlds were, for example, well-known to Buddhism. Thus, in Buddhist writings, often cast in allegorical or poetic form, we learn of: a series of heavens in three divisions; of gods inhabiting the palaces of the Sun, Moon, and stars; of a 'ten thousand world system'. In the *Lotus of the True Law* Buddha addresses an assembly of wise men from so many worlds that 'they numbered eight times the grains

5   *Plato: prestigious anti-pluralist of the fourth century BC.*

Plato.

of sand in the river Ganges'; he spoke of 'atoms of earth of fifty hundred thousand myriads of kotis of worlds' and after similar multiplications asks of his audience 'if any one should be able to imagine, weigh, count, or determine [the number of] those worlds?' Significantly, Buddha included amongst his 'fourteen difficult questions' which were ultimately fruitless whether or not the world and the self were finite, infinite, eternal or not eternal (or both, or neither).

Pluralism of a mildly speculative kind was also present in classical China. Teng Mu, a sage of the thirteenth century, wrote:

> Empty space is like a kingdom, and heaven and Earth are no more than a single person in that kingdom. Upon a tree there are many fruits... how unreasonable to suppose that besides the heaven and Earth... there are no other heavens and other Earth!

A pluralism of analogy and probability which echoes the sentiments of Metrodorus, and anticipates those of Tsiolkovsky — as well as their very imagery.

Hinduism was also well placed to develop pluralism to which it was sympathetic: 'The Indian philosophical and cosmological framework... explicitly assumes the existence of extraterrestrial intelligences.'[9] The Vedic writings allude to a planetary hierarchy living under Brahma and containing 400,000 humanoid races, differing in their appearances and cultures, although able to work together. A contemporary Indian mystic[10] has written of 'unlimited planets in the spiritual sky' related to 'material planets' in a proportion of 3:1, of spiritual interplanetary travel, and that 'we are only a fractional portion of all the living entities in the many universes of the material world'. But, unaccompanied by the requisite scientific enquiry, the otherwise amenable culture of India did not nurture a pluralism capable of secular elaboration

Some civilisations with considerable astronomical skills bequeathed little which might be interpreted as pluralism: the mathematically accomplished Maya, the great Polynesian navigators, the Chaldeans, and the megalith builders of Northern Europe. Pluralism only arose when certain necessary conditions coincided: where, for example, astronomy and at least a rudimentary cosmology have developed, and where questions of ultimate origins and purposes arise. Its emergence probably required good, clear skies at night, a settled polity, an economic surplus to support intellectuals and their speculations — and some prestige accorded to intellectual courage and originality.

Here and there in the ancient world, from the wine-dark Aegean to the Yellow Sea, pluralism made its appearance, most remarkably in Greece and its dependencies. But then it faded with the decline of the classical world. After a promising start it was to live in the cultural margins for well over a thousand years.

References

1   R. Thiel, *Und Es Ward Licht*, Hamburg 1956; translated as *And There Was Light*, New York, 1957. Thiel suggested, however, that Auaximander postulated a spherical, as opposed to a cylindrical Earth.

2   F. M. Cornford, 'Innumerable Worlds in Pre-Socratic Philosophy', *Classical Quarterly*, 28, 1934 — formidable scholarship which suggests Anaximander had in mind a succession of worlds, not a set of co-existing worlds.

3   Or did he? See F. M. Cornford, *op cit*.

4   R Thiel, *op cit*; J. L. E. Dreyer, *A History of Astronomy from Thales to Kepler*, 1906.

5   The concept of successive worlds has also been credited to Anaximander; see F. M. Cornford, *op cit*.

6   Petron's work is an example of the tenuous nature of evidence for classical pluralism; it is known from Plutarch's writing which drew in turn on Phanias of Ereson, reporting on Hippys of Rhegium.

7   For example in Stanisław Lem's metaphysical novel *Solaris* (1961) and Tarkovsky's film of the same name (1971).

8   B. Pullman, *The Atom in the History of Thought*, 1998.

9   Ronald Huntington, 'Mything the Point: ETIs in a Hindu and Buddhist Context' in James Christian (ed) *Extra Terrestrial Intelligence...* 1976. Although in Walter Sullivan, *We Are Not Alone*, 1964; revised 1993, we learn of Hindus that 'Their various cosmologies do not envisage other inhabited worlds.'

10 A. C. Bhaktivedanata Swami Prabhupada, *Easy Journeys to Other Planets*, 1970.

3 Medieval pluralism

After Rome: pluralism in the shadows

For the first thousand years of the Christian era pluralism was in abeyance in the West, together with much of the particular Greek cosmology of which it was a part. But when it revived, from c1200 it flourished suddenly, splendidly and in novel ways, a desert bloom emerging after a span of unkind aridity.

During the years of its quiescence the lands of Islam became the chief repository of the Greek intellectual achievement. Although the Arabs preserved and refined Greek astronomy and cosmology they were not overly interested in pluralism. In Europe conditions were actually inimical to the idea. The confusion and disruption attendant on the fall of the Roman Empire, an uninquiring intellectual climate, and the hostility of the early Christian church towards the pagan cosmology of the Greeks, left the remnants of classical pluralism in a hostile environment.

Early Christian comments on cosmology had to square with the Scriptures, for example that the heavens were stretched out as a curtain, or framed as a tent — the 'tabernacle theory' of the cosmos. Another bibilical source of cosmological confusion, then and later, was the concept of a 'firmament in the midst of the waters' — were there, therefore, waters above and below the Earth? These metaphors and morphologies were incompatible with sophisticated Greek notions of a spherical Earth and an inhabited antipodes. Isidore of Seville (c570-636) and The Venerable Bede (673-c735) were amongst the first commentators to refer again to a spherical Earth, and the concept was widely reinstated by the mid-ninth century.

Pluralism was not entirely ignored during this scientifically unproductive epoch, but the treatment it received did nothing to advance it. The most authoritative statement on the subject came from St Augustine (354–430) who, in his *City of God* firmly opposed pluralism in either sense: as a succession of worlds or as a simultaneous plurality of worlds.

Although pluralism seemed to slumber in the first millennium, appearances were deceptive. Interest in pluralism has been encouraged in more modern times by technological innovation: the telescope, or artificial satellites. In the Middle Ages it was an indirect beneficiary of technological and economic developments[1] which made possible a slow but perceptible growth in population: the horse harness and deep plough; the rotation of crops; wind- and water-powered grain mills and an economic surplus which supported intellectuals, and the building of monasteries and great cathedrals. Universities founded in the high Middle Ages, especially Paris and Oxford, were to be particularly important to pluralism, as were the monastic orders, notably the Dominicans and Franciscans. Within this framework the pluralists and anti-pluralists had time and space to work and think.

Pluralism in the later Middle Ages

The ice age of pluralism came to an end in the later Middle Ages, when the lost works of the Greeks became available, particularly from Arab sources. Aristotle's *De Caelo* (6) was translated by Gerard of Cremona and was read once more from *c*1170. This and other Aristotelian works enjoyed authority and prestige, so that their author's anti-pluralism influenced and persuaded the first wave of philosophers who produced commentaries in the thirteenth century. Thus Roger Bacon (*c*1220–92) generally esteemed for his spiky intellectual independence was a slightly disappointing anti-pluralist, as was William of Auvergne (*c*1180–1249). In addition to Aristotle's arguments these philosophers produced another: a plurality of worlds would imply immense and unacceptable voids between world-systems; there could not be such a 'nothingness' in God's creation.

St Thomas Aquinas (1224–74), well versed in Aristotle, passed judgement against pluralism in his voluminous works. In his *Summa Theologica* he opposed it on the grounds that if other worlds were similar to this one then they would be pointless, surplus to requirements; if dissimilar 'none could contain all things' (as this world does, according to Aristotle) 'and would therefore be imperfect, and so not the work of the Creator'. The 'uniqueness' of this world did not compromise the concept of the omnipotence of God. This latter proposition is significant; it was in terms of a possible — and unacceptable — limitation of the power of God that the supporters of reborn pluralism were to rest their case. The potential for tension between Christian doctrine and Aristotelianism was to become nowhere more clear than in pluralist discourse.

Pluralism's attractions were intoxicating to the scholars of the high Middle Ages. Aquinas's own tutor and fellow-Dominican, Albertus Magnus (*c*1205–80) admitted:

> Since one of the most wondrous and noble questions is whether there is one world or many... it seems desirable for us to enquire about it.

Although the initial enquiries resulted in negative answers, closely derived from Aristotle, important changes were imminent as Christian theology, once a serious obstacle to pluralism, became the unexpected catalyst in its revival.

The scholastic debates during the thirteenth century repeatedly encountered the contradiction between Aristotle's geocentric denial of pluralism, favoured on the whole by the academics in the universities, and the religious doubts about this held by the academics of the Church. Since the University of Paris, the major centre of European learning at the time, was a seat of such controversy, the matter was put by some theologians in 1277 to the Bishop of Paris, Etienne Tempier, for a definitive ruling. The outcome was a turning-point in the history of pluralism, possibly of modern science.

6   *The medieval cosmos: or is it? A famous engraving depicting the pre-Copernican Earth as centre of the Universe with its cosmic machinery lying beyond the sphere of the fixed stars; modern scholarship suggests that Camille Flammarion, nineteenth century astronomer and pluralist (and accomplished artist and engraver) may have been its originator.*

Tempier condemned, on papal authority, 219 propositions, many derived from Aristotle. Amongst them (proposition 34) was the notion, implied by Aristotlean science 'that the First Cause [God] cannot make many worlds'. The ruling did not state that God had necessarily caused or wrought many worlds, merely that He could if so minded.

This opened the floodgates for a critical scrutiny of Aristotle's denial of pluralism. Jean Buridan, rector of the University of Paris, tackled Aristotle's chief argument against pluralism, the notion that each element (fire, air, water, earth) had its natural place *vis–à–vis* this world, and no other world. Surely, argued Buridan, an all-powerful God could create other elements with their own natural places *vis–à–vis* other worlds? In commenting on *De Caelo*, Buridan concluded: 'Just as God made this world, so He could make another or several worlds.'

The English Franciscan, William of Occam criticised the Aristotelian propositions with similar subtlety. Would not a flame burning in Oxford (his University) 'tend towards its own place', that is Oxford, and a flame in Paris do the same? And yet, if those flames were swapped about they related to their new homes; in Oxford a Paris flame would not lean towards Paris. Similarly, earth (that is, soil, matter) in our world fell towards our Earth; on another world it would drop towards that other Earth. Aristotle's assertion that the doctrine of natural place necessarily refuted pluralism, apparently contained inherent contradictions. William of Occam also addressed the question: 'Can God make anything better than He has made?'[2] He concluded that whilst God could make this world better in 'essential goodness', He could make another world like ours but separate from it, better in 'accidental goodness'. He also developed St Augustine's idea that God could create a 'perfect man' — if man, then species; if species, then world containing that other, better species.

William of Occam's philosophical convolutions, set as they were in the medieval debate on the relative merits of Aristotelian cosmology and Christian theology, may strike the impatient modern mind as so much hair-splitting. They were important within the paradigm of their day and assisted the vigorous renewal of pluralism even before the Renaissance. He explored realms of philosophy and ontology which barely engage with contemporary discourse but were then significant. For example, he advanced arguments about whether the qualities of individual entities can be apprehended by resemblances between things (nominalism) or by reference to universal qualities (universals). This related directly to pluralism: if there were more than one world, then entities might not be able to share common universals; if different worlds already shared universals then God could only create new worlds by adjusting all existing worlds so that they shared the same universals.

By using pluralism as a weapon in their broader arguments the theologians and philosophers developed it in new and unusual ways. For some authorities this was ominously subversive of cherished Aristotelian ideas. Oxford University declined to award William of Occam a Master of Theology degree; the long tradition of anti-pluralist intolerance was alive and well again.

Pluralism continued to grow in strength, nevertheless. The tutor to the future king, Charles V of France, Nicolas Oresme (1325-82) pursued a similar line to that of Buridan and Occam, but he took it further. He felt that there was no reason for denying a plurality of worlds either in time (the Stoic argument) or place. God could, if He wished, create other worlds like and unlike our own, by virtue of His omnipotence. Like Occam he concluded that bodies moved about with reference to their particular surroundings and not to a single, cosmic centre. For example, does not a stick of wood that might fall in air rise if dipped in water? Oresme was careful to keep well within the bounds of Christian orthodoxy; omnipotent God was 'infinite in His immensity' and so, if more than one world existed 'no one of them would be outside Him or His power'. He added that in spite of this divine omnipotence 'of course, there never has been nor will there be more than one corporeal world...' Oresme was not the last philosopher to hedge his bets in this way, to assert pluralism as a possibility but to express doubts as to its actuality.

Towards the end of the medieval contribution to pluralist theory, however, came the first intimation of a different and more potent proposition. In 1440, Nikolaus Krebs (Cardinal Nicholas of Cusa, 1401-64) published *De Docta Ignorantia* (Of Learned Ignorance) in which he took a bold line:

> Rather than think that so many stars and parts of the heavens are uninhabited...
> we will suppose that in every region there are inhabitants, differing in nature by
> rank and all owing their origin to God...

In so saying, Nicholas of Cusa took pluralism beyond mere speculation about the
existence or non-existence of other worlds to thoughts about the inhabitability of those
worlds, even about the kind of beings to be found in them. The classical world had started
this particular line of speculation in a modest way; Nicholas of Cusa launched it on a
course which has since grown in importance to the point where it has become the major
element of contemporary pluralism, concerned closely with the nature of extraterrestrial
intelligence and the means of detecting it. Nicholas of Cusa hazarded a few guesses about
the ETs 'of worlds other than our own [of whom] we know still less, having no standards
by which to appraise them'. It might be, for example, that inhabitants of the solar regions
might be 'bright and enlightened denizens... more spiritual than such as may inhabit the
Moon, who are possibly lunatics...'.

Although late medieval pluralism had developed far, its position was far from secure. As
in classical Greece it remained a minority interest. It was a subordinate part of larger
debates; it continued to have its critics; it pulled its punches — as the case of Oresme
demonstrates — and it existed largely within commentaries rather than as a fully
developed set of propositions. By dwelling on a plurality of 'worlds' as such, and (*pace*
Nicholas of Cusa) declining to discuss the nature and characteristics of those worlds, in
particular their inhabitants — if any — it remained on relatively safe ground. Once actual
or possible extraterrestrial beings entered the discussion it would take an ominous turn.
Christianity rested on the assumption that humanity was corrupt and fallen, and that God
had made Himself incarnate, in the person of Jesus, by way of offering redemption to the
fallen inhabitants of this Earth. But what then of other Earths, of other *kosmoi* should they
exist?

At the very end of medieval pluralism the French theologian William Vorilong,
commenting on Aristotelian cosmology, looked back to the scholastic concern about limits
to the power of God, and forwards to the novel question of extraterrestrial life. He
suggested in the old style: 'infinite worlds, more perfect than this one, lie hid in the mind
of God.' On the other hand he also enquired about people on another possible world:
'whether they have sinned as Adam sinned' and whether the death of Christ would
redeem any fallen extraterrestrials. Vorilong thought that the Crucifixion might have
redeemed them: 'Christ is able to do this even if the worlds were infinite' but he added 'it
would not be fitting for Him to go to another world that He must die again.' This was the
first straw in the wind of a complex debate which was to colour and inflame pluralism for
at least two centuries. Its echoes are with us still.

Shortly afterwards, in 1473, the *De Rerum Natura* of Lucretius was published, having
been rediscovered in 1417. Whilst this central work of classical atomism supplied fruitful
sources for the pluralists it also complicated their task. Atomism was less easy to reconcile
with Christianity than the geocentric, implicitly anthropocentric, Aristotelian doctrine
which sat more easily with Christian beliefs about the special, central place of humanity

in God's schemes. There is, of course, no logically necessary contradiction between Christianity and pluralism, but to certain casts of the theological mind there can be; this contingency was about to have its field day.

Like its classical predecessor, late medieval pluralism lay within a metaphysical framework. The 'naked eye astronomy' which had existed since time immemorial provided some fundamental facts: that there was a cosmos, a Sun, a Moon, stars, and planets. Pluralism pursued its course without further reference to specific astronomical data. That too was about to change drastically.

References
1   Lynn Whyte Jnr, *Medieval Technology and Social Change*, 1962
2   William of Occam was here commenting on the *Book of Sentences* of Peter Lombard (1095-1165).

4 Renaissance and the new cosmology, 1540-1640

The three keys

After many centuries of quiet and neglect, pluralism returned dramatically with the Renaissance. In the last years of the Middle Ages commentators had already introduced one of the fresh elements that was to hasten and inform its growth: the potent concept of extraterrestrial life. Three major developments brought pluralism (**7**) back into the arena: a theory, a device, and a discovery.

The *theory* was that of Nicholas Copernicus who stated that the Earth orbited the Sun and not vice-versa. This hypothesis explained a great deal of astronomical difficulty simply and neatly; it dealt a mortal blow to the Ptolemaic system and raised crucial questions about the size, shape and disposition of the cosmos. Geocentric objections to pluralism were scattered. This revolutionary theory generated strong opposition and took a long time to spread. In modern times, occasional public opinion polls suggest that pre-Copernican ideas about the cosmos are still far from dead, a commentary, perhaps, on the efficacy of public education.

The *device* was the telescope (**8**), invented in the first decade of the seventeenth century, possibly earlier. This was the first major technological boost for pluralism, not to be equalled in its impact until the development of the spectroscope in the 1860s.

The *discovery* was by Europeans of the Americas during the period 1490-1520. Others had been there before: the Vikings, possibly people from Ireland, China and Japan. The Native Americans had been there for millennia in any case. No matter, to the European cast of mind during the Renaissance this was a dramatic discovery, full of implications for a pluralism which had begun to speculate about extraterrestrial beings. A papal bull of 1537 declared Native Americans to be human; if fellow-humans had been encountered in the 'Antipodes' what of other worlds?

Of all these developments, the invention of the telescope by Hans Lippershey (1608)[1] and its almost immediate use by Galileo was arguably the greatest catalyst for pluralism. Galileo published his first findings in 1610. The telescope revealed myriads of faint and hitherto unknown stars, suggesting that the doctrine of equidistant 'fixed stars', welded to a revolving sphere, might be faulty. The Moon was (in Galileo's words) 'full of inequalities, uneven, full of hollows and protuberances, just like the surface of the Earth itself...' Most pregnant of all, on 7-8 January 1610 Galileo discovered four moons moving around the planet Jupiter. At a stroke the uniqueness of the Earth as the centre of the cosmos and the only possessor of a satellite was undone; the potential for pluralism of this single, practical discovery was enormous.

But the first blow to geocentrism had been struck over seventy years before with the publication in 1543 of *De Revolutionibus Orbium Coelestium* (On the Revolutions of the Heavenly Orbs), in which Copernicus demonstrated that the Earth was not the centre of cosmos, rather that it orbited the Sun (**9**). Copernicus also stated the distance from the Earth to the Sun was small compared to the distance from the Earth to 'the height of the firmament' — the realm of the fixed stars. The configuration and scale of the cosmos were radically altered.[2]

Kepler, Bruno and the English bishops

Another input to the cosmological revolution, the work of Johann Kepler (1571-1630), particularly his *New Astronomy* (1609) explained the geometry and dynamics of the orbits of the planets around the Sun. John Donne expressed (1611) the growing bewilderment at the loss of the old cosmology:

> *The Sun is lost, and th'Earth and no man's wit*
> *Can well direct him where to look for it.*
> *And freely men confess that this world's spent,*
> *When in the Planets and the Firmanent*
> *They seek so many new...*

The attitudes of the three scientists: Copernicus, Galileo, and Kepler to a plurality of worlds are indicative of its gradually changing fortunes in this period of intellectual change. Copernicus was quiet on the subject; Galileo was uncharacteristically cautious, even ambiguous, but Kepler was a pluralist enthusiast who added a contribution to the science fictional exegesis of the subject in the manner of Lucian, and other later scientists.

It was not, however, the reactions of the pioneers to a plurality of worlds that heightened interest in the doctrine but rather the comments of their critics and supporters. It is sometimes claimed that the 'dethronement' of the Earth from its central, cosmic position accounted for the increasing hostility of the Catholic Church towards the new cosmology. 'Ever since Copernicus', thought Nietzsche, humanity had been slipping 'away from the centre into the unknown'. But the extraordinary fact of the Earth being peripheral in the cosmos, contingent upon the Sun, may, however, have been of less importance in the long run than the notion that humanity might, after all, not be unique or uniquely important to the Creator, in whose image, the Bible stated, it had been created.

When reborn atomism reinforced the Copernican cosmos, theologians perceived a really serious threat from pluralism. The resulting controversy was heated and lengthy. By no means all of the the theological critics were Catholic. The protestant Philip Melanchthon wrote in 1550 (significantly, in a science textbook) that the notion of a plurality of worlds was 'absurd'; it offended both science (Melanchthon still accepted the Aristotelian assumptions) and theology — Christ was sent to redeem fallen humanity on

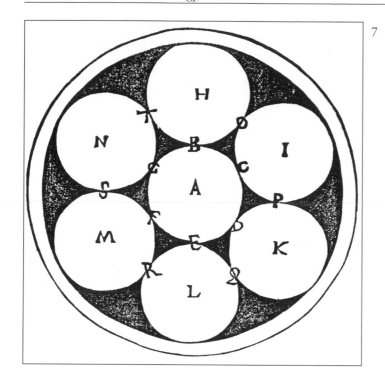

Giordano Bruno's many worlds: a set of Copernican systems supporting a plurality of worlds; from De Immenso et Innumerabilis, *1591.*

this Earth alone and there was no warrant for assuming that He was sent to other inhabited worlds. It was unthinkable that putative inhabitants of other worlds were naturally perfect. Since they could attain salvation in no other way than divine Incarnation and that arrangement was unique to Earth, there could be no extraterrestrial beings.

The debate was quickly engaged throughout Europe. In England, Thomas Digges[3] translated a part of *De Revolutionibus* (1576) subtitling his book: 'A Perfit Description of the Caelestiall Orbes According to Most Aunciente Doctrine of the Pythagoreans Lately Revived by Copernicus...'. By claiming continuity with Pythagoras (as had Copernicus) Digges may have hoped to allay fears about the new order, but the undisguised references to pluralism in the foreword to this translation may have had the opposite effect. The French essayist Montaigne commented (1580) on the pluralism which now burgeoned in the light of Copernican theory, but with a Gallic detachment that left crucial questions open. To Montaigne, Copernicus may or may not have been right; similarly it might be a delusion to 'make the Moon a celestial Earth', but on the other hand eminent thinkers, past and present, support the notion so perhaps it was plausible after all. If, as Democritus, Epicurus and others held 'there be many worlds... how do we know if the principles and laws of this one apply to the others?'

One notable philosopher and scientist, however, seized upon the pluralist potential of the new, heliocentric cosmos and worked it vigorously: the Italian Giordano Bruno (7). Although his response was early, nevertheless it remained a model example of ultra-hard line pluralism. Bruno, an enthusiastic Copernican, lectured on the new cosmology in Oxford (1583). His pluralism drew on Copernicus, on atomism — he was well-versed in Lucretius — and on the revived doctrine of plenitude which was under discussion again

8   *A plurality of worlds?*
Heroic representation of
the archetypical
Renaissance astronomer
from J.P.Nichol's
Architecture of the
Heavens, *1856.*
Nichol, a popular writer
who lectured in Britain
and America was an
ardent pluralist; to him
the Orion nebula held
'germs' of 'every varying
degree of intelligence,
and every shade of moral
sensibility and
greatness'.

in the late Middle Ages. Bruno's views were set down in three works: the dialogues *La Cena et le Ceneri* and *De l'Infinito Universo* (both of 1584), and the poem, *De Immenso* (1591). Amongst his *obiter dicta* one finds:

> 'We join world to world and star to star in this vast ethereal bosom' [ie, not as described in Aristotle's sealed and limited cosmos];

> 'Why should infinite amplitude be frustrated, the possibility of an infinity of worlds be defrauded?'

Bruno stated (in *De l'Infinito Universo*) that the Universe was filled with numberless worlds 'all placed at certain intervals from one another' and that both stars and planets could be homes to extraterrestrial beings. He was the first advocate of 'high pluralism', an extremist tendency which is inclined to populate any, even every, celestial object. He was also among the few pluralists who have speculated that these 'other worlds' — stars, planets, even comets and meteors — might possess souls, divinely inspired animation of their own; panpsychism. An infinity of worlds; inhabited worlds; animate worlds: pluralism could hardly go further. For all his extremism, Bruno prevaricated when it came

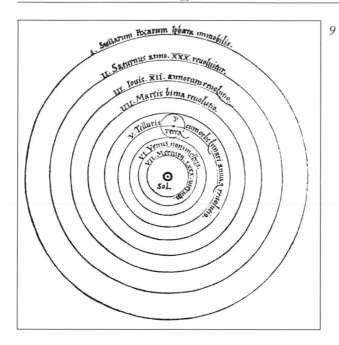

9　*The Copernican system, with the Sun at its centre thus demoting the Earth and removing a major barrier to pluralist thought; from* De Revolutionibus Orbium Coelestium, *1543.*

to consideration of the difficult theological questions posed earlier by Melanchthon and Vorilong: do the inhabitants of these other worlds live in a state of grace; has God made Himself incarnate on other worlds? Bruno tended to evade the full implications of the problem by generalising about the essential unity of the cosmos, the similarity of its components, the goodness of God.

Bruno was tried as a heretic, and burned at the stake (1600) — not, as is occasionally asserted, because of his pluralism but (amongst other heresies) for denying the divinity of Christ. His unremitting advocacy of pluralism nevertheless did him no good in the eyes of Church authorities who were becoming increasingly touchy about the implications of the new cosmology and scientifically-based aspects of pluralism.

Galileo, who was also to suffer for his association with the 'New Science', was more circumspect than Bruno. Although pluralism was a live issue, Galileo said nothing about it in his major work *Sidereus Nuncius* (1610). In his next work, however, *Istoria e Dimonstrazioni Intorno alle Macchie Solari* (1613) he wrote of the 'false and damnable view of those who would put inhabitants on Jupiter, Venus, Saturn and the Moon, meaning by 'inhabitants' animals like ours, and men in particular.' He added, somewhat cautiously, that these celestial bodies might, or might not, be inhabited by entirely different creatures, 'I should... neither affirm nor deny it'. By 1632, possibly because of the official pressures on him, he had retreated slightly from even this position.[4] He played safe: there might be entities on the Moon far different from anything we could imagine; if so these beings would praise their Creator for ever.

If Galileo remained ambiguous about pluralism, his fellow-countryman Tomasso Campanella did not. Campanella, a monk who had incurred the displeasure of the Jesuits and the Inquisition, wrote his defence of Galileo from a Naples prison, an arrangement which itself begs some interesting questions. In his influential *Apologia pro Galileo* (1622) Campanella tried to square Galileo's work and Copernican astronomy, in particular their pluralist implications, with the Bible. For example, Galileo's early suggestion that he had observed water on the Moon is compared to *Genesis* 1, verses 6-7 'And God made the firmament and divided the waters which were under the firmament from the waters which were above the firmament.' Campanella also reasoned that even if the Bible did not mention a plurality of worlds as such, it did not deny or forbid it either — not convincing proof but an argument which kept the arena open and was to be employed again frequently. In spite of his ingenuity Campanella was no more successful than his contemporaries in reconciling the concept of extraterrestrial living beings with Christian orthodoxy, in particular with the Incarnation. He tried various tacks: if extraterrestrials were not descended from Adam they might be free of sin; and again, Christ had 'united Heaven and Earth with God the Father.'[5]

Whilst the pluralists and their critics debated, massive disruptions in Europe dwarfed these intellectual concerns, but nevertheless affected them. This was an era of furious religious strife between the Protestant Reformation and the Catholic Counter-Reformation, notably in the vastly destructive Thirty Years War (1618-48). Although there was no strict causal relationship between this or that faith and pluralism, in general Protestantism was more accommodating towards it than Catholicism. The pattern was often complex. Kepler, a protestant working often within Catholic Europe gave rather more aid and comfort to pluralism than the Catholic Galileo or his bolder exegete, Campanella. This was possibly because the regimes within which the Catholics worked were more sensitive about theological and scientific threats to orthodoxy and had the official apparatus and concepts (the Holy Inquisition and 'heresy') with which to combat them.

Kepler's pluralism grew from his refining and extending of Copernicanism; it seems to have drawn little, if at all, on classical or late medieval thoughts about the principle of plenitude, the inevitable plurality of worlds arising either from God's infinite power, or other large-scale characteristics of the cosmos like the infinity of atoms. He had observed a 'new star', a supernova in the constellation Ophiucus, the Serpent Bearer, in 1604. Two years later he described this phenomenon in *De Stella Nova*, advancing the pluralist case on the analogic grounds that the ideas of Copernicus 'admitted the Earth to the citizenship of the heavens' and that all stars, including his supernova, were composed of the same material as the Earth. Since this material had brought about life on Earth, it might well do so elsewhere. In a letter of 1607[5] Kepler wrote that he agreed with Bruno 'that the stars are inhabited' adding: 'I am the more ready to adopt this view because I believe with Aristarchus [therefore with Copernicus, as his correspondent pointedly responded] 'that the Earth moves just like the planets.'

But Kepler, enmeshed in theological and cultural questions, entertained doubts about the implications of pluralism; a bolder pluralist than Galileo, but a switherer at this stage nonetheless. In *De Stella Nova* he made a special plea for humanity on this Earth: possibly it had not been spontaneously created out of some universal 'vital faculty' of the other celestial bodies, but had been created especially by God. Had God created the new star as a beacon for all His creation? Kepler had no doubts about there being numerous creatures elsewhere in the Universe, able to gaze upon his new star and to wonder about it; to some of them it might 'be more amenable to their understanding than to ours'. This was one of the first authoritative suggestions that humans might not be the lords of creation, but might share the Universe not only with other beings but with superior beings. It was one of the most disturbing questions raised by pluralism and one which was to haunt Kepler — and humanity — thereafter.

Kepler deflected the full implications of his question by employing his astrometrical skill: it appeared that the inhabitants of other worlds would not actually be in a position to see the new star, or the coincident and remarkable conjunction of Mars, Jupiter and Saturn; human dignity and uniqueness were saved, if only just. Kepler's keen pluralism found another and less hazardous outlet, one already employed by Plutarch (whose pluralist work he knew) and a form frequently used for similar reasons by scientists in modern times: science fiction. Kepler was a rare phenomenon, a first-rate scientist with a splendid literary imagination; in his scientific work pluralist inclinations were hedged about as their author kept an eye on the powerful and occasionally threatening ranks of his critics. In fiction he could speculate more easily. Even so, his imaginary voyage to the Moon, *Somnium* (The Dream) drafted in 1609 was only circulated as a manuscript; it was not published until after his death in 1634.[7]

Although *Somnium* has been acclaimed as '*fons et origo* of the new genre'[8] of fictional Moon voyages, even of space-fiction, the book rested heavily on the scientific credentials and astronomical work of its author, proclaimed on the title page to be 'Imperial Mathematician', Kepler's official position. The free imagination of the author only applied to the beginning and end of the book; its heart is a clear description of the Moon's topography, all the more remarkable for having been written before Galileo observed it with his telescope.[9]

Somnium is the story of a young Icelandic scientist who is told in a dream about the Moon and its inhabitants by a spirit. The story is based upon Copernican astronomy; the Moon's inhabitants gaze upon a revolving Earth (hence, named *Volva*), not a fixed Aristotelian one. One half of the Moon, 'Subvolva' always faces the Earth; the other, 'Privolva' faces away; there are complex — and astronomically correct — observations of the Earth from the Moon and the clear implication that the Moon-dwellers have the intelligence to make such observations and a range of related deductions. What of these beings? Kepler suggests that since the mountains and valleys of the Moon are larger than those of Earth, the lunarians are also larger than us; they grow more quickly, have larger appetites and die sooner than Earth-dwellers. The massive swings of the lunar climate cause the 'Privolvans' on the dark side of the Moon to be nomadic, also they take to water easily by way of cooling themselves, and — precursor of much later science fiction including that of H.G.Wells[10] they live much of the time in sublunarian caves. The Earth-facing 'Subvolvans' are more settled and urbanised. Although Kepler skated near to the crucial question of the possible humanity and divinely ordained-origins of the lunarians, he managed to avoid too close a contact with the issue. *Somnium* is the most dramatic pluralist literary speculation of the period of the Renaissance and Reformation, more measured, economically expressed and complete as a literary work than, for example, Bruno's effusions.

The spirit and purpose of *Somnium* had a close rival in an English work: *The Discovery of a World in the Moone* (published anonymously, 1638) by the Rev John Wilkins, brother-in-law of Oliver Cromwell, and later Bishop of Chester. Wilkins also used his book to advance a serious scientific case. His closest English progenitor was Nicholas Hill, a philosopher and pioneer of the atomist revival whose *Philosophia Epicurea* (1601) discusses the plurality of worlds, drawing on Democritus and Copernicus. Hill's thought exemplified the secular temper of atomist-based pluralism; a plurality of worlds which arises not so much from 'plenitude' and the omnipotence of God, but rather from the logic of a Universe composed of a single substance in atomic form from which our Earth has been fashioned to bear living creatures.

By analogy, the other heavenly bodies must be formed and populated similarly. Hill made some allowances for local variation: Moon-dwellers were smaller, and Sun-dwellers larger, than Earth-dwellers by virtue of the different sizes of their worlds. But, like many of his contemporaries, Hill found the theological implications of many worlds to be an exacting problem to which he had no convincing answer; he fell back on assertions about the special status of Earth-dwellers over all others in the cosmos.

English intellectuals were becoming well aquainted with continental pluralist debate; Hill was one of a group containing John Donne, Sir Walter Raleigh, and Christopher Marlowe. The Rev John Wilkins, a convinced Copernican, was much influenced by Campanella's defence of Galileo. In his *Discovery* he set forth fourteen pluralist propositions which include:

1 That the Moone may be a world.

2 That a plurality of worlds doth not contradict any principle of Reason or Faith.

5 That the Moone hath not any Light of her own.

9 That there be high Mountains, deep vallies, and spacious plains in the body of the Moone.

13 That 'tis probable there may be inhabitants in this but of what kinde they are, is uncertain.

14 That 'tis possible for some of our posteritie, to find out a conveyance to this other world; and if there be inhabitants there to have commerce with them.

The last, most prescient proposition was added to the third edition of his work (1640) after Wilkins had read a fictional account of a trip to the Moon, *The Man in the Moone: or, A Discourse of a Voyage Thither...* written *c*1628 by Francis Godwin, another Anglican bishop, but published posthumously a few months after Wilkins' own book (**10**). Although Godwin's work shows clearly that pluralism was very much 'in the air' in late Jacobean England it was not, like Wilkins' *Discovery*, a scientific treatise.

The *Discovery* was published with the imprimatur of the Bishop of London. It stirred up a furious controversy, not in the least because Wilkins declined to take a literal interpretation of each and every passage in the Bible. One of his antagonists, Alexander Ross, expressed his anger in the title of his rejoinder: *The New Planet No Planet: or, the Earth No Wandring Star; except in the Wandring Heads of Galileans.* To Wilkins, pluralism was a 'new truth' in the sense that it was becoming scientifically based, dealing in such discoveries as

10 *Off to the Moon: propelled by birds and a sail, 'Domingo Gonsales' heads spacewards in Bishop Godwin's* The Man in the Moone *(1638).*

the work of Copernicus and Galileo, and the observed mountains and valleys on the Moon; he did not draw on medieval concepts of divine omnipotence and plenitude. But he had to face the same intractable problem as his contemporaries: if there were extraterrestrial beings (which he thought probable) what was their status with regard to sin, grace and redemption? Wilkins raised the issue only to veer away leaving it to others to decide; he speculated that other-worldly beings might be very different from Earth-bound humans in any case.

Wilkins' *Discovery* is a fitting emblem for pluralism at the end of the Copernican century, *c*1540–1640. It demonstrates how far the concept of the plurality of worlds had developed, along what lines, and at what cost. It had moved on from its late medieval and essentially religious, theological basis with an emphasis on the omnipotence of God, to 'early modern' foundations which emphasised the possibility, even probability, of there being other worlds, perhaps inhabited worlds, because observation, analysis, and the new cosmology led to such a conclusion. Some writers attempted a syncretic solution, combining elements of theological and secular pluralism. They were to continue doing so for a long time; traces of this attempted synthesis are still current. From whatever standpoint the Christian pluralists proceeded, they now faced the difficult issue of the spiritual qualities of extraterrestrial beings, their possible status as fallen or redeemed beings, and the especially searching question of the means of that redemption. In short, had God made Himself incarnate on more than one world, and if so, how?

In spite of its dramatic advances, pluralism remained a minority interest; many intellectuals, particularly clerics, were sceptical or hostile. Pluralist debates barely reached the majority of the population. The response of some major thinkers and literary figures was also decidedly underwhelming: Shakespeare had little to say about it, although he did employ the theme of the 'music of the spheres' from time to time. Milton was to make a few glancing references only.[11] To Pascal a plurality of worlds that might 'know nothing of us' merely added to the 'infinite terror' he felt when contemplating 'the eternal silence of those infinite spaces'.

Nevertheless, the pluralist bequest of the Renaissance was of inestimable importance in the history of the long debate about life on other worlds. It introduced new, essential concepts and started the process of establishing arguments on a scientific basis.

References
1   But who was the begetter of the spyglass? See H. C. King, *The History of the Telescope*, 1955.
2   On Copernicus, see T. S. Kuhn, *The Copernican Revolution*; E Rosen, *Three Copernican Treatises*.
3   Amongst the first to suggest determining stellar distance by parallax, not accomplished for some 350 years thereafter.
4   *Dialogio dei Due Massimi Sistemi del Mondo*.
5   *Ephesians* 1, 10; *Colossians* 1, 20.
6   To J. G. Brengger, a physician of Kaufbeuren who had criticised this reasoning.
7   English translation, *Kepler's Somnium* by Edward Rosen, 1967.
8   Marjorie Hope Nicolson, *Voyages to the Moon*, 1948.
9   But was Galileo the first so to observe the Moon? The Englishman Thomas Harriot may have predated Galileo; even if he did, his work was little known and unlikely to have come to Kepler's attention.
10 Cf, H. G. Wells, *First Men in the Moon*, 1901.
11 See utterances of the Archangel Raphael, *Paradise Lost*, 7, lines 621-2.
12 For an alternative view, that the burgeoning of pluralism was essentially philosophical and metaphysical rather than sprung by science and technology, see Arthur O. Lovejoy, *The Great Chain of Being*, 1936

5 Baroque perplexity, 1640-1740

During the century from 1640 the advance of pluralism accelerated. There were about seven books devoted to the plurality of worlds published 1584–1684, but seventeen 1657–1757[1]. The geographical focus of pluralism began to shift; for much of the period French thinkers dominated the scene. England's rapid and forceful entry into the debate reflected her growing wealth, the development of science, and the formulation by English thinkers of a 'natural theology' which at last gave some positive religious encouragement to pluralists. The growth of English science was reflected in the foundation of the Royal Society (1662), the first secretary of which was the author of *The Discovery of a New World*, the pluralist Rev John Wilkins, about to become Bishop of Chester. Thenceforth the English-speaking world took a leading place in pluralist discourse: first in England alone, then later in partnership with Scotland whose contributions were remarkable and disproportionate to the size of her population; later still the United States, now the pre-eminent source of pluralist theory and practice.

By the mid-eighteenth century pluralism had settled into a more comfortable position within European culture; not quite an orthodoxy, but far removed from being a near-heresy, perceived as a threat to established power, and dangerous to advocate. The opening years of 'baroque pluralism' were, however, marked by neutrality and caution; tentative assertions, with escape clauses.

Pluralism received two encouragements: the revival of atomism and the work of René Descartes — himself a pluralist fence-sitter. The revival of atomism cut both ways; it encouraged pluralism but, because of its atheist stripe, it also encouraged hostility to it. The neo-atomists, like their classical ancestors Epicurus, Lucretius and Democritus, suggested not only that an infinity of worlds might be possible but, more controversially, that the Universe might be a self-governing mechanism which acted irrespective of divine cause or intervention. Some neo-atomists attempted to engage God in some way with their systems, but the problem of self-regulation remained inherent in atomism and was hard to avoid.

Other theological questions remained. What was the moral and religious status of the denizens of other worlds? What did the existence of these beings imply for humanity on Earth, which might no longer be unique; the apple of God's eye? The pluralists moved into a conceptual labyrinth involving cosmic vortices, universal gravity, revived atomism and implicit criticism of Christian orthodoxy. In doing so they displayed considerable originality and courage; little wonder that some of them hedged their propositions or took refuge in ambiguities.

Doubts in high places: Descartes and Newton

The cosmology of René, Descartes which encouraged the pluralists was enshrined particularly in his *Principia Philosophiae* (1644). This drew on atomist assumptions in its suggestion that the Universe was composed of an infinity of Copernican systems, such as the one in which the Earth was situated. Each system was a 'vortex' (*tourbillon*) which bordered on others in an infinity of 'vortices'(**11**). Each star was a potential replica of the Sun, attended by planets.

Descartes distanced himself from the pluralist implications inherent in his vortices. In private correspondence he stated that he 'did not infer from all this that there would be intelligent creatures in stars or elsewhere' although, on the other hand 'I also do not see any reason by which to prove that there were not'; he preferred to leave the question 'undecided'. In the *Principia Philosophiae* Descartes made no specific mention of other worlds or their inhabitants, although he doubted that God had created 'all things solely for us'.

Towards the end of his life Descartes became a cautious, closet pluralist. He ventured an opinion to the Dutch philosopher Frans Burman: 'An infinite number of other creatures far superior to us may exist elsewhere.' He was being unusually bold here: the possibility of a plurality of inhabited worlds was a hazardous enough concept; to hint that the inhabitants of these worlds might be in some way superior to humanity which enjoyed a 'special relationship' with God bordered on heresy. As it was, the works of Descartes were placed by the Catholic Church on its Index of prohibited books shortly after his death; his caution about the pluralist question was not misplaced.

Isaac Newton also played a cautious hand. His cosmological framework was ostensibly less encouraging to pluralism than the Cartesian vortices. Nonetheless, by advancing the prestige of science, it encouraged the spirit of objective enquiry upon which pluralism eventually came to depend. Newton himself said little about the subject and might have said even less were it not for the promptings of the Rev Richard Bentley, later Master of Trinity College, Cambridge, who was anxious to reconcile Newton's master-work, the *Principia Mathematica*, with revealed religion. Although unravelling Newton's assertions on pluralism from Bentley's interpretations of them is not straightforward, it is clear that whilst Newton had no hostility to pluralism as such, he was adamant that God alone had created the cosmos and that it was not subject to structural change. He carried this argument forward in the second edition of *Principia* (1713) in which he opined that if the stars were 'the centres of other like systems, these, being formed by the like wise counsel, must all be subject to the dominion of One'. He thought also that 'in God's house (which is the universe) are many mansions' and asked 'why should all these immense spaces of heavens above the clouds be incapable of inhabitants?'[2] Although lacking positive support for pluralism, these open-minded expressions of pluralist possibilities from an illustrious source did pluralism no harm.

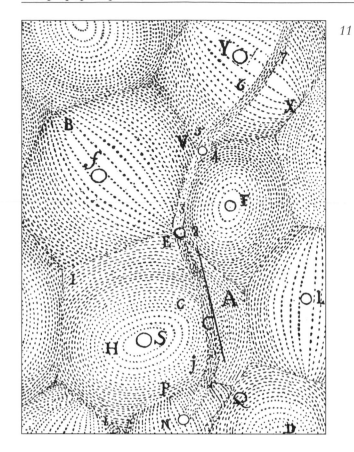

11 *Cartesian cosmology: some of the* tourbillons *of René Descartes, a Universe composed of an infinity of vortices, therefore of a potential multiplicity of worlds; from* Principia Philosophiae, *1644.*

The debate widens: Gassendi, von Guericke, and the Cambridge School

There were others who took a bolder line, usually with due care paid to religious sensibilities. Pierre Gassendi, a French priest and mathematics professor, set out to reconcile atomism with Christianity and in so doing touched on the vexations of pluralism. He posited God as 'first cause', creator of the atoms and thereby setter-up of the cosmos.[3] But he rejected the original atomist idea of *aperoi kosmoi*, an infinity of worlds. Gassendi was well-versed in the medieval scholastic arguments about the principle of plenitude. He commented that God might or might not have created more than one cosmos; there was no hard proof either way and unless there was evidence to the contrary, this 'world' (cosmos, world-system) was unique.

But what of other worlds like the Earth within our 'world' or cosmos? Gassendi was prudent. Whereas the Moon and the planets were most certainly not the residences of beings in any way superior to humans (for this would smack of heresy), he wondered if there might be 'men, or certain forms of life... similar to men' elsewhere in the heavens. Taking his cue from the biological diversity of our planet, he concluded there might be entirely different forms of life, for example on the Moon. This remained an unproved theory however, as was his speculation that the nearer to the Sun was an inhabited globe, the more perfect would be its inhabitants; most noble of all would be the sun-dwellers themselves, if they existed. Although he raised potent questions about the possibility and purpose of other inhabited solar systems elsewhere in the Universe, he fell back on arguments about the unknowability of such issues for mere humans. This style, of implicit and explicit conditionals, of ifs and maybes, characterised other contemporary commentators on pluralism.

The Protestant scientist Otto von Guericke, burgomaster of Magdeburg and pioneer of vacuum technology, set out his own arguments in the *Experimenta Nova* (1672) wherein he also described the 'Magdeburg hemispheres' experiment which demonstrated the existence of atmospheric pressure. Guericke rejected the notion of 'Aristotelian' mini-universes, but like Gassendi he felt that there might be a plurality of worlds like the Earth, even of solar systems surrounding stars other than the Sun. Using the principle of plenitude, and in an attempt to reconcile pluralism with Christian doctrine he added: 'A plurality of created worlds not only does not detract from the majesty... of the Deity but in fact makes it more resplendent'. Also, like Gassendi, he favoured the idea of other worlds 'formed with such diversity' that each could be distinguished from each 'in many thousand ways'.

Guericke was ambivalent about lunar life; if there were creatures on the Moon then they would be 'of a baser kind' than humans, but the Moon 'lacks animate life' because of its harsh conditions. He thought the planets might bear life and threw doubt on the religious idea that they were created, like the heavens in general, for the convenience of humans on this planet. And yet they might have been created for the convenience of other creatures, but if so these would be unlike humans who were 'the pinnacle of creation'. But then again, Guericke wondered if God had revealed some of His secrets to 'other', and therefore implicitly superior, creatures elsewhere in the Universe. In so speculating, he prodded the mine which lay at the heart of baroque pluralist debate, the question of the divinely-ordained uniqueness and superiority of humanity — although by using the considerable possibilities of rhetoric, he avoided prodding it too obviously.

English Protestant thinkers also juggled with pluralist questions in the light of Cartesian thought, notably Henry More of Cambridge. Being set in verse, his *Democritus Platonissans, or, An Essay upon the Infinity of Worlds out of Platonick Principles* (1646) could excite the consciousness and take liberties in ways denied to strictly scientific and theological prose. More presented a robust line on pluralism, but he knew the weight of the opposition. He started his book thus: 'Infinitie of worlds! A thing monstrous if assented to, and to be startled at.' Having read his Descartes carefully he observed perceptively that the French philosopher, the 'sublime and subtil mechanick' as he called him, had temporized on pluralism: 'though he seem to mince it must hold infinitude of worlds, or which is as harsh, one infinite one'. Anxious to reconcile Epicurean atomism with aspects of Plato's philosophy, More was able to employ the pluralist implications of Descartes' vortices: what had been done in 'this Terrestial starre' had also been done 'in every orb beside'. The stars were 'innumerable numbers of fair lamps' orbited by other worlds 'Whereof the number I deem infinite.'

More linked plurality in time, the 'succession of worlds' with the newly-emergent Copernican line of 'simultaneous worlds': 'To weet that long ago there Earths have been/Peopled with men and beasts before this Earth.' But, although More held to his pluralism, he gradually abandoned the philosophical foundations upon which he had placed it, eventually rejecting the work of both Descartes and Epicurus. Addressing the question of reconciling pluralism with the Incarnation, More was amongst the first to suggest that the problem could be solved if God had made known the Incarnation and Crucifixion on this Earth to other worlds beyond. This idea was to have a long career, and still has its supporters.

More's Cartesian pluralism influenced other English theoreticians, notably Robert Boyle and Ralph Cudworth whose work *The True Intellectual System of the Universe* (1678) spoke of 'this dull earth of ours' being a planet in the 'vortex' of the Sun, and that many people have 'shrewdly suspected that there are other habitable globes besides this earth'.

Popular pluralism in France

Whilst theoreticians and philosophers worked out the pluralist implications of Cartesian cosmology, two French writers, Cyrano de Bergerac and Bernard le Bovier de Fontenelle, offered popular pluralism to the educated public, the first clear example of what was to become a major feature of its development. Both men sensed acutely the considerable market awaiting them. They were the first to realise that pluralism, by virtue of the mystery, wonder and scope of its subject is almost bound to appeal to an enquiring laity.

Cyrano de Bergerac entered the fray in a breezy style with two novels[4] — one about a journey to the Moon and the other to the Sun, usually treated as one work, *L'Autre Monde* or *Histoire Comique* (1657-62) (**12**). He had partly learned his pluralism from Gassendi who had once tutored him; he was sympathetic to an atomism which did not find it necessary to call upon divine creation or intervention. His extreme audacity in this and other questions reflected his reputation as one of France's most skilful swordsmen and duellists, accomplishments by which he is better known to posterity. In the Moon novel, which starts by offering the reader a strong dose of Copernican cosmology, the traveller (**13**) finds there the Garden of Eden, giant people and animals — implying that the Moon people are of a higher order than animals and as noble as people on Earth; a contention calculated to rile Church authorities. This line is pursued further when the lunar priests doubt the traveller's rationality. There is a similar putting-down of anthropic pretensions in the Sun novel which is strongly suffused with atomist themes: the spontaneous generation of life and living people from heated mud, and the repeated, spectacular, re-forming of living beings. The Sun is represented as a home for souls from all other worlds; a secular heaven.

Cyrano's text presented complex suggestions at three levels: a straight story of space fiction; an encounter with exotic aliens; a set of questions undermining anthropocentrism; and a more general scepticism about latent, possibly universal ethnocentrism. Whilst advancing pluralism wittily and readably, it demonstrated, as Lucian had in classical times and tens of thousands of future narratives were to do, the fathomless opportunities for imaginative fiction presented by the pluralist idea.

A more measured and yet equally witty and informative excursion into fictional pluralism, emerged in 1686, de Fontenelle's *Entretiens sur la Pluralité des Mondes*

12   *Title page of the first English edition of Cyrano de Bergerac's* Histoire Comique *(1687).*

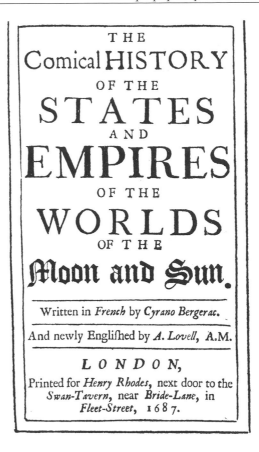

THE
Comical HISTORY
OF THE
STATES
AND
EMPIRES
OF THE
WORLDS
OF THE
𝕸𝖔𝖔𝖓 𝖆𝖓𝖉 𝕾𝖚𝖓.

Written in *French* by *Cyrano Bergerac.*

And newly Englifhed by *A. Lovell,* A.M.

LONDON,
Printed for *Henry Rhodes,* next door to the
Swan-Tavern, near *Bride-Lane,* in
Fleet-Street, 1687.

('Conversations on the Plurality of Worlds') the first book to argue the pluralist case popularly (**14**). Within a year it was censored by being placed on the Index[5], not that this seems to have curtailed its popularity; it was reprinted 33 times during the author's own lifetime. It was to have some 30 English printings in seven separate translations, three within two years of its publication; it was also translated into nine other European languages. Its argument is framed around a series of dialogues between a polished philosopher-scientist and a quick-witted aristocrat, 'the Marquise', the first woman to enter the pluralist debate fully and publicly. The fact that she was fictional is of less account than the well-attested knowledge that educated women read and discussed the *Entretiens*. Joseph Addison noted in the *Guardian* of 8 September 1713 that Englishwomen 'of quality and fortune', inspired by Fontenelle's writings were discussing pluralism and Copernicanism; although he also added facetiously: 'dividing their speculations between jellies and stars'.

In spite of Fontenelle's serious intentions and his care to ground pluralist dialogue in contemporary science, such as Descartes' vortices, it was his occasional flights of fancy that often caught readers' imaginations. Thus the Marquise suggested that the inhabitants of Venus, being closer to the Sun than those of the Earth, were 'little black people, scorched with the Sun, witty, full of Fire, very amorous...'. By comparison, on Jupiter, the Jovians were dull and heavy-going.

13 *Cyrano de Bergerac en route to the Moon in a glass bubble heated by the Sun; lunar inhabitants, flora and fauna aloft; from* A Comical History of the States and Empires of the Moon, *1687 (translated from the French original of 1657).*

Fontenelle exercised discretion by stating that these various extraterrestrials were not humans. This circumspection was clearly insufficient for the Church authorities who sensed danger in Fontenelle's pluralism by analogy: if a world is like the Earth in certain respects, why not in others? Why should not its inhabitants be like humans? Fontenelle implied that the extraterrestrials were very much like humans: some of them even gazed at the Earth through telescopes; the Marquise echoed the tastes of contemporary French aristocracy when she surmised that the Venusians 'invent masques and tournaments in honour of their mistresses'.

Although Fontenelle was plausible and ingenious, he was unable to solve the abiding problem of how to free the idea of a plurality of worlds from the charge that it undermined humanity's special place in the Universe. Even so, he probably did more to raise the consciousness of educated men and women about the plurality of worlds than any other contemporary, perhaps of any writer between classical times and the great Victorian popularisers.

The lesser-known French pluralist work, *Traité de l'Infini Créé*, (1769, but composed earlier) probably by the Abbé Jean Terrasson, offered a heady draught of high pluralism influenced by Descartes and Fontenelle. The *Traité*, took a hard, idiosyncratic line on the sensitive question of pluralism and the Incarnation. It suggested that God had become Incarnate on the multitude of planets inhabited by humans in an infinite succession of vortices, in space and time, and that angels were the survivors of past planets. But was this a serious thesis or an early satire on pluralist optimism and the theological coils into which pluralist discussions could fall? Opinions have long been divided on the matter. At least the existence of an early pluralist parody suggests considerable popular, even fashionable, acquaintance with pluralist ideas, implicit homage to Fontenelle and Cyrano de Bergerac.

Pluralism plainly stated: Huygens

The Dutch scientist Christiaan Huygens, one of Europe's leading astronomers and physicists, had long been sympathetic to pluralism before he wrote *Cosmothereos*, posthumously published in 1698 and for a long time a primer of pluralism. It vied with Fontenelle's work which it resembled in important respects. The basis of Huygens' pluralism lay in the propositions which Fontenelle had already made: that other planets resembled the Earth; that they were made to bear life because no other purpose was conceivable for them; that the further a planet was distant from the Sun the more night-lighting moons its inhabitants enjoyed by way of compensation; that nature was bounteous and that one had therefore to assume plurality in the absence of good arguments to the contrary. He drew on his astronomical work to suggest that the Moon was probably not inhabited but, by stretching the interpretation of his observations, he thought that the presence of clouds and water could be detected on Venus and Jupiter. He asked why it should not be that, beyond the solar system, the stars 'of the same nature as the Sun' should not have in turn their own planets with their own moons. Although he offered an optimistic line, Huygens respected scientific method and was careful to present these conjectures as such, and to employ rhetorical questions instead of assertions when raising unusual possibilities. Huygens' work lay the ground for the rational, common-sense pluralism of the Enlightenment. Unlike Fontenelle's fiery Venusians and lethargic Saturnians, Huygens' extraterrestrials were very much like Earth-people having similar mathematics, music, geometry and astronomy; similar powers of reasoning and similar moral sensibilities endowed by a just, equal-dealing and beneficent Creator. The suggestion that the human type was the standard cosmic model offered the attractive implication of a universality of human reason. Although it was to have its critics, this line of argument remained influential throughout the Enlightenment and strong traces of it were to endure longer. They lie at the heart of contemporary theories of how we might communicate with extraterrestrials regarding, for example a shared comprehension of binary mathematics presumed to exist throughout the Universe.

Natural theology: pluralism as a prop to religion

Huygens's *Cosmothereos*, originally set in Latin, went into some half-dozen English translations as well as four other European languages. Armed with this, and Fontenelle's *Entretiens*, educated society in the early eighteenth century, particularly in Northern Europe, was becoming familiar with the concept of pluralism.

14   Entretiens sur la Pluralité
des Mondes: *de Fontenelle's
philosopher and the Marquise
discuss the plurality of worlds
in the heavens above them
(1687), an illustration from
the epoch-making pluralist
text.*

Richard Bentley, who corresponded with Isaac Newton on the subject of other worlds, was the first Newtonian to push the implications of pluralism further than more orthodox contemporaries felt prudent; certainly much further than Newton felt wise or necessary. In a series of eight lectures intended to prove the truth of Christianity (1692)[6] Bentley argued that the fixed stars which resembled the Sun might have orbiting planets, although he was careful to add that the soul of a virtuous man outweighed the worth of all heavenly bodies in the sight of God. As the Earth was created for humanity, so the heavenly bodies might have been created for other intelligent minds. In skating around the theological difficulties of pluralism Bentley entered equally dangerous territory: he suggested that God might have created a range of scientific laws which could operate differently on different worlds, an ingenious notion although a curious one for a disciple of Newton. He postulated also that God might have created a range of extraterrestrials, of 'Orders and Classes' of minds, quite different from humanity. Since these beings constituted 'different species' the Incarnation might not then apply to these other worlds. At first glance Bentley's omnipotent, law-changing God might only have extended the medieval concept of plenitude, updated for the expanding Universe of the Newtonians. In so doing Bentley advanced the concept most resisted by many other Christian theologians, the dread thought that humanity might not be uniquely chosen, or even gifted with superior powers.

Thus, from the late seventeenth century, pluralism was palpably 'in the air', appearing with increasing frequency in scientific and philosophical works, particularly in England. It was endorsed by the pioneer vegetable anatomist Nehemiah Grew (1641-1712) and the 'father of English natural history' Thomas Burnet (1635-1715), friend of Bishop Wilkins. The philosophers John Locke (1632-1704) and Bishop George Berkeley (1685-1753) supported it. In Locke's *Essay Concerning Human Understanding* extraterrestrials may have capacities which transcend the limitations of human senses. In *Alciphron* (1732) Berkeley ventured into territory that had given continental Catholics so many problems: it may be that fallen humanity is condemned to live on the planet Earth which is an exception amongst the myriad of worlds, possibly the dungeon of the cosmos in which, elsewhere 'there are innumerable orders of intelligent beings more happy and more perfect than man...'

Other European philosophers commented upon or developed pluralism. Gottfried Leibnitz (1646-1716) who had long dwelt on the issues raised by pluralism wrote in his *Theodicy* (1710): 'Today... it must be acknowledged that there is an infinite number of globes... which have as much right as it [our world] to hold rational inhabitants, though it follows not at all that they are human...' The German polymath declined, however, to amplify the theological implications of his pluralist assertions. On the other hand, another German, Christian Wolff (1674-1754) a frequent advocate of pluralism, committed himself on the nature of extraterrestrials, in particular the inhabitants of Jupiter. Since less sunlight reaches Jupiter than the Earth, Wolff's Jovians had big eyes, and since the eye and body are proportionate, Jovians are, he calculated pedantically, probably over 13 feet tall.[7]

It was, however, in England that the main pluralist developments were taking place. The Rev William Derham (1657-1735), chaplain to the future George II, published the third book (after those of Huygens and Fontenelle) which was to underpin later Enlightenment pluralism: *Astro-Theology, or a Demonstration of the Being and Attributes of God from a Survey of the Heavens* (1714). It tapped the new-found popular interest in the plurality of worlds, the subject which opened his book and introduced its main arguments. It remained in print for much of the century; there were 14 English editions alone and it was a major vehicle for introducing popular pluralism to Germany where it had six editions — the first in 1732. Derham's book was a pioneer of the school of 'natural theology' which provided a congenial basis for religious faith in the Enlightenment: the proposition that the existence of God could be inferred from studying the natural order; through biology, physics or astronomy for example. Faith arising from revelation was to become for many polite thinkers suspect as 'enthusiasm', not entirely *comme il faut*. Far from grating on religious sensibility, pluralism was now employed — particularly in Protestant Europe — to support, even to demonstrate, the existence of God through the application of reason. Derham was a high pluralist; he populated the Moon and asserted 'there are many other Systemes of Suns and Planets besides that in which we have our residence'. Like Huygens, Derham was intrigued by the teleological aspects of pluralism: what was the purpose of creating other worlds? Derham did not think that the whole Universe was created solely for humanity but, unlike Huygens, and also unlike Fontenelle, he deemed it fruitless to speculate about the actual nature of extraterrestrials; the reader is referred to *Cosmothereos* should he or she wish to guess about 'the Furniture of the Planets ... what Vegetables are produced ... what animals live there ... what endowments they have...'

Alexander Pope encapsulated the new pluralist mood in his *Essay on Man* (1733) which included the oft-quoted lines:

> *He, who thro' vast immensity can pierce,*
> *See worlds on worlds compose one universe,*
> *Observe how system into system runs,*
> *What other planets circle other suns,*
> *What vary'd Being peoples ev'ry star...*

These were possibly inspired by a lecture given by the astronomer William Whiston, successor to Newton as Lucasian Professor of Mathematics at Cambridge, another high pluralist who speculated that beings might not only live on, but even within, stars, planets and comets.

Whiston's lecture had been arranged by Joseph Addison of the *Spectator*, widely read by commercial and fashionable London society and beyond. It often pressed a pluralist line, for example by endorsing Fontenelle's work and by arguing (1712) that every star performed 'the same offices to its dependent Planets, that our glorious Sun does to this...'.

By 1740 pluralism was fairly well established amongst the European, and particularly North European, intelligentsia. It was becoming a common topic of conversation amongst the men and women of 'polite society', the subject of whole books, as well as sections or paragraphs in other ones, of essays, articles, and poetry. It benefited from the spread of science, particularly the works of Copernicus, Descartes, and Newton who, even if they had not addressed the issue directly, nevertheless supplied crucial frameworks and data for pluralist arguments.

There remained the unresolved questions that kept old arguments alive and set new ones in train. Was humanity unique? Were extraterrestrial beings like humans, better or worse, brighter or duller? How did a plurality of worlds, and of inhabited worlds, fit into God's plans? The deeply difficult question to Christians of God's Incarnation on one planet, the Earth, burned still, although its fires were banked in the era of rational piety and Deism, particularly in the salons of Protestant or anti-clerical intellectuals. The foundations for pluralism's contribution to the Age of Reason were well laid.

References

1 Bibliography of pluralist books published before 1917, Appendix in Michael J. Crowe, *The Extraterrestrial Life Debate* 1750-1900.

2 This passage exists in different forms, some truncated; nevertheless, it is indicative of Newton's prudent line on pluralism.

3 In his posthumous *Syntagma Philosophicum*, 1658.

4 *Les Estats et Empires de la Lune*, 1657; and *Les Estats et Empires du Soleil*, 1662.

5 Not for the last time; it went back on the Index in 1900 having been released in 1825.

6 Later published in part as *The Folly and Unreasonableness of Atheism*.

7 Vol III of *Elementa Matheseos Universae* of 1735.

6 The light and shadow of reason, 1740-1800

Encyclopedism and mysticism

Pluralism has had affinity with a wide range of religious, ideological and cultural forms. In 'The Age of Reason' many pluralists believed that their case rested on rational, logical propositions. Through much of the eighteenth century, however, pluralists also employed the time-honoured methods and assumptions of analogy and probability in one form or another rather than demonstrable, reasoned proof of a 'plurality of worlds' verified by astronomical observation.

Until the late eighteenth century, the mass of pluralist writers still felt it necessary to reconcile pluralism with the claims of religion. They were to continue to do so for many years but by 1800 the first deliberately secular essays in pluralism were appearing. Tom Paine (1737–1809) thought pluralism a good stick with which to beat organised religion. It was also to be deployed as a social or political critique, for example in the struggle against slavery.

In 1795, however, the Royal Society published a paper by the German-British astronomer William Herschel[1] in which he claimed that his pluralist deductions came from 'astronomical principles', not from the 'fancy' of theology or poetry. But Herschel was a high pluralist; was he really 'deducing' from observations, or was he finding evidence to shore up *a priori* assumptions?

Another factor which appeared to be at odds with the notion of a rational, objective pluralism was the appearance of pluralist mysticism during this period, notably the religious ideas of Emanuel Swedenborg which depended heavily on the idea of life on other worlds. This was a significant historical development: from this time forwards pluralism gave aid and comfort to a wide range of religious and secular beliefs and cults. These two aspects of pluralism, its rational and mystical wings, have long been in tension.

The French encyclopedists, the embodiment of detached, reasoned enquiry, inclined favourably towards pluralism. The article on Epicureanism in the *Encyclopédie* waxed enthusiastic about the pluralist implications of atomism; it found 'nothing repugnant' in the concept of pluralism, 'worlds similar to ours, as well as different, can exist'. The writer of the article, probably Denis Diderot drew on Cartesian forms to expand the point; other worlds were 'great vortices' filling 'the infinity of space'. Diderot had put the pluralist case previously[2] by stating 'when matter... brought this world into being, creatures like myself were of very common occurrence. But might not worlds too be in the same case?' Similarly, Diderot's fellow-encyclopedist Jean Le R. d'Alembert contributed an article on

An Inhabitant of the DOG STAR.

15   *An alien from the age of reason: George Cruikshank's imagined Sirian for a later edition of* The Travels and Surprising Adventures of Baron Munchausen *(1786).*

cosmology that supported the theory of a chain of being, an hierarchical universe in which each star supported a retinue of planets. D'Alembert was a critical pluralist; elsewhere he admitted that by no means all planets, still less comets, were candidates for habitation. Another encyclopedist, Baron d'Holbach, although a keen pluralist, cast doubts on the anthropocentric assumptions suffusing many pluralist arguments. He sounded a warning note about the presumed humanoid forms extraterrestrials might take since the conditions of other planets might differ greatly from those of Earth. This issue still lies near the heart of pluralist debate, for example in the form of anthropic and other 'chauvinisms'.

Two other French philosophers contributed to the longer-term development of pluralism, although in strongly contrasting ways. The Comte de Buffon's massive *Histoire*

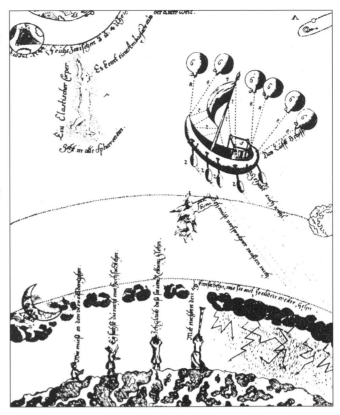

16   *Spacecraft of the Enlightenment: held aloft by evacuated copper globes and propelled by paddles from Earth to a German-speaking Mars; an illustration from Eberhard Kindermann's* Die geschwinde Reise... *of 1744, well laced with technical detail drawn from astronomy and the imagined botany and zoology of other worlds.*

Naturelle (1749 onwards) suggested that wherever the same conditions prevailed, particularly temperatures, similar life-forms would emerge. He elaborated this hypothesis by calculating the past, present, and future temperatures of planets, then estimating their life-bearing spans. Thus life would appear on the 'second satellite of Jupiter' 196,266 years hence, and would last for 206,602 years; on Mars and the Moon life had already run its course. Buffon's remarkable work presaged the evolutionary pluralism of a century later.

The Abbé Condillac, on the other hand, sounded a timely warning against the analogical reasoning of the pluralists. In *La Logique* (1780) he analysed pluralist methodology and demonstrated its refinements and weaknesses. For example, merely to assert that other planets are inhabited simply because Earth is, represents a weaker form of reasoning than suggesting that planets with similar qualities to Earth (for example, having diurnal and annual revolutions) might resemble Earth in other ways. Neither analogy constitutes proof, however; a point to be taken up by later logicians, particularly John Stuart Mill.

There were many other pluralists amongst the intellectual adornments of the age of reason; a sensible pluralism was practically one of the desiderata of a right-thinking *savant*. Benjamin Franklin, for example entered in his 'Articles of Belief' of 1728, a personal position statement: 'I conceive it [space] fill'd with Suns like ours, each with a Chorus of Worlds...'. Nearly 30 years later he consoled the readers of his 1757 almanac with the thought that if the Earth were to be annihilated by a comet 'it would scarce be miss'd in the Universe' since 'There are an infinite Number of Worlds under the Divine Government.' Franklin's aside: 'We must not presume too much our own importance' touched on a sensitive and well-established issue. Lord Bolingbroke's posthumous *Philosophical Works* (1754) addressed it rhetorically by suggesting that one might easily imagine creatures 'tempered in a finer clay, cast in nobler moulds, than the human' and, more audaciously 'animated by spirits more subtle and volatile than ours'.

Pluralism was able to support with equal facility the enthusiastic and up-beat Franklin and the world-weary Tory aristocrat who thought, on contemplating the plurality of worlds, that every being in the Universe 'is adapted... to the place he is to inhabit'. He might as well be content with that position since 'the general state of mankind... is a state not only tolerable, but happy', pluralism as ever telling us more about ourselves than about extraterrestrials.

Voltaire exploited pluralism skilfully. His own thoughts about it were measured and sceptical; he was generally in favour but was suspicious of hard-line interpretations such as Huygens' notion that there were creatures living in comets. His pluralism had also to take some of the strain arising from his abandonment of Cartesian 'vortices' in favour of Newton's conceptually less favourable cosmology. Although he was not the first writer to employ the device of an extraterrestrial passing comments on the curious ways of humanity, he used this device with great skill, avoiding tediously predictable angles. In his *Micromégas* (1752) the eponymous hero (a Sirian 120,000 feet in height) travels about space encountering inhabitants of various planets including Earth-bound humanity whose self-important pretensions he duly deflates (**15**).

In contrast, *Die geschwinde Reise auf dem Lufft-Schiff nach den obern Welt...* ('The Swift Journey by Airship to the Upper World...' 1744, revised 1784) by the German astronomer Eberhard Kindermann (**16**) was within the 'great tradition' of space-fiction like Wilkins' *Discovery* of a century before. Kindermann had already pressed pluralist arguments in his textbook *Vollständige Astronomie* ('Complete Astronomy', 1744) in which data of astrometry (distances and orbits of planets, etc) sat side-by-side with descriptions of life and conditions on, for example Mars or the Moon. His underlying theory was that the more splendid a planet appeared, the more advanced were its inhabitants. In the *Swift Journey* five cosmonauts travel to a 'moon of Mars' (a prescient guess since no Martian satellites were known until 1877) where they find happy and advanced extraterrestrials living in a state of grace; beings who admire the spaceship in which their guests have come, as an indication of human development and sophistication.

Whereas the equation of technological with moral progress was an optimistic Enlightenment conceit, unconvincing to later generations, the idea that some extraterrestrials were either free from sin or were otherwise superior to human beings was neither new nor peculiar to the age. It had the attraction of being able to curb religious pretensions whilst at the same time offering hope of the possibility of actual moral development. Pluralism could also be enlisted in the struggle against slavery and racism. David Rittenhouse, an American astronomer and a popular public speaker remarked with complex obliquity (1775) that 'our globe' was 'but an inconsiderable part of the whole. Many other and very various orders... may, and probably do exist, in the unlimited regions of space.' Some of these 'orders' may have their existing happiness increased by being cut

17 *'The eye of Providence'*
stares out from the many
worlds of Thomas Wright of
Durham in his Original
Theory... *(1750).*

off from the Earth so that he could say of them: 'We have neither corrupted you with our vices nor injured you by violence. None of your Sons or Daughters... have been doomed to endless slavery merely because their bodies... reflect or absorb the rays of light, in a way different from ours....'

The connections between pluralism and astronomy in the eighteenth century, however, were generally more straightforward than the metaphors of Kindermann or the liberal orations of Rittenhouse. A large number of astronomers were openly avowed pluralists, although few of them actually supported their pluralist views on the interpretation of observations. In three remarkable instances pluralism may have been prior to, or at the least an effective cause of, astronomical enterprise: Thomas Wright, Immanuel Kant, Pierre Laplace.

Thomas Wright of Durham (1711-1786), a keen pluralist, had two highly original cosmological ideas well ahead of their time (**17**). He was not, however, an astronomer's astronomer but rather a religious mystic with astronomy as his chosen interest. In his *An Original Theory or New Hypothesis of the Universe* (1750) he advanced the remarkable and correct hypothesis that the Milky Way is an optical effect, a view he later revised, however.[3] In amplifying his idea he went off-beam with notions of it being a ring of stars affixed to a sphere, or perhaps an annular construct like the ring of Saturn. This perceptual breakthrough was not, however, Wright's chief concern. Much of his *Original Theory* is concerned with the plurality of worlds about which he wrote enthusiastically: 'when we consider them [stars] all as flaming Suns... *Primum mobiles* of a still much greater Number of peopled Worlds, what less than an Infinity can circumscribe them... or less than an Omnipotence produce and support them?' God had created a 'Universe of Worlds, all deck'd with Mountains, Lakes, and Seas, Herbs, Animals...'. Wright speculated that in this Universe, where 170 million inhabited worlds lay within humanity's 'finite view': 'Man may be of a very inferior class; the second, third or fourth perhaps...'.

The starry heavens of Kant and Herschel

Wright's principal route to later fame was a brief review of his *Original Theory* read by Immanuel Kant who, at this stage in his career, was working towards a publication on cosmology: *Allgemeine Naturschifte und Theorie des Himmels* ('General Natural History and Theory of the Heavens', 1755). Inspired by Wright's thoughts he adapted them and added two original ideas of his own. The *Allgemeine Naturschifte* first brought together three mainstays of modern cosmology: that the Milky Way was the edge of a disc viewed from within; that the nebulae are similar but distant constructs[4], and that planets are formed out of nebulae surrounding stars.

These original thoughts on galactic topology have sometimes been rather grudgingly recognised by the astronomical fraternity — like Wright, Kant was not an astronomers' astronomer. Although Kant's cosmology arose from the thoughts of a supreme specialist in the categorisation of perceptions, his pluralism was grounded in traditional methodology: simple analogy in which stars are so many Suns 'centres of similar systems in which all may be arranged... as in ours'; and the plenitude of a 'Godhead' the field of whose 'divine attributes' was as infinite as their manifestations.

Kant imagined that 'Millions and entire mountains of millions of centuries will flow by, within which always new worlds and world-orders form themselves...'. These worlds form a hierarchy in which the 'most perfect classes of rational beings' were further from the cosmic centre. Hence, the inhabitants of Mercury were stupid, those of Earth 'exactly the middle rung' on the ascent to the brightly gifted and morally excellent extraterrestrials living on Jupiter and Saturn. This was the reverse of Wright's hierarchy in which imperfect beings dwelt in the far reaches of the cosmos whilst 'final perfection' lay at its centre, the chain of planets between the two being a long series of locations for purging, rewarding and perfecting souls as they made the journey from the periphery to the 'Centre of Gravitation', the throne of God. Kant elaborated his theory by positing humanity as an 'intermediate state', living in a region where sin is possible but from which the souls of the dead can move forth to more perfect worlds.

Kant's well-known statement in his *Critique of Practical Reason* (1788) that the two things which 'fill the mind with ever-increasing admiration and awe' the more they are reflected on: 'the starry heavens above me and the moral law within me' is more compelling when

18   *Sir William Herschel, discoverer of Uranus, leading astronomer of his day — and an enthusiastic pluralist, amongst the minority who suggested an inhabited Sun, 'a most magnificent habitable globe', like the stars themselves.*

perceived against the background of his pluralist cosmology. His 'starry heavens' were pluralist heavens: so many world-systems set in an orderly, hierarchical framework; a grand and rational plan, divinely ordained.

Like atomism, the Kantian cosmology was congruent with pluralism although not linked to it by logical necessity. Later in life Kant, still a convinced pluralist, distanced himself from the more speculative flights of pluralist imagination in the *Allgemeine Naturschifte*. He was not alone in finding inspiration from pluralism for his wider work. Professional astronomers often found pluralism congenial and inspiring.

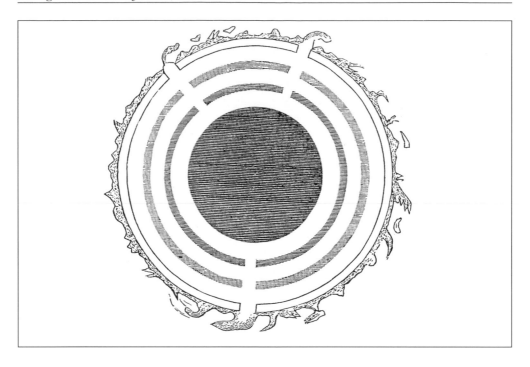

19   Herschel's inhabitable Sun, its cool, inhabitable body protected from the fiery surface by two insulating layers, occasionally penetrated by holes (sun spots) which reveal the darker, inner globe. Other advocates of solar life included Gauss, Dick, Brewster, Flammarion, and Colonel Woodward Warder (1901) whose Sun 'bore peerless cities of smiling grandeur, inhabited by noble denizens of angelic goodness'.

Roger Long, appointed Lowndean Professor of Astronomy at Cambridge in 1750 suggested[5] that the solar system had been organised by God so that the hotter, inner planets were cooled by mines of saltpetre whilst the colder, outer ones were warmed by 'subterraneous fires'; he declined, however to commit himself as to the precise nature of their inhabitants.

Two better-known French contemporaries of Roger Long also advocated pluralism: Lalande and Laplace. Jerome Lalande was the author of the standard French textbook on astronomy, the *Traité d'Astronomie* (1792) a sober work which presented a pluralism derived from analogy: since the Earth was made for habitation 'one cannot refuse to admit that the planets were made for the same purpose'. More revealingly, Lalande thought 'Imagination pierces beyond the telescope; it sees a multitude of worlds infinitely larger....' By way of encouraging this imagination, Lalande published a new edition of Fontenelle's *Entretiens* in 1800.

Pierre Laplace, the leading mathematical astronomer of the late eighteenth century, linked pluralism to his influential nebular hypothesis: that the solar system originated from the condensation of a nebula rotating the Sun. This theory has had a chequered history. Now back in favour it was in decline particularly in the 1920s and 1930s which were also a thin time for pluralism. Laplace discussed pluralism and his nebular theory in the same, last, chapter of his *Exposition du Système du Monde* (1796). His pluralism, like that of Kant or Lalande was based largely on analogy. He was cautious in writing about extraterrestrials whom, he supposed, were varied in forms which were probably determined by ambient temperatures. Having presented the nebular hypothesis, Laplace added that the many stars of the Universe are so many suns 'which may be the foci of planetary systems'.

Two other astronomers were pre-eminent as committed pluralists, however: Johann Bode, director of Berlin Observatory, and the versatile William Herschel, Court Astronomer to George III and discoverer of the planet Uranus in 1781 (**18**). Both astronomers pressed the case for an inhabited Sun (**19**), the certain mark of a high pluralist. Like Giordano Bruno three centuries earlier, Bode wrote of inhabited stars, planets, even comets. To those who doubted the possibility of such extreme pluralism, Bode responded that 'rational inhabitants, and even the animals, plants etc, of other planetary bodies' were simply formed differently from human beings. Like many other advocates of extreme pluralism, he employed the rhetorical interrogative: 'Who can conceive of what special arrangements of the wise Creator in regard to the climate, zones... sectioning of creatures... may not be expected for all those on a cometary body?'

William Herschel was no less keen but he played the pluralist line far more cautiously. Early on in his astronomical career, and probably inspired by the Scottish pluralist James Ferguson, he confided in his private notes that it was 'beyond doubt' that the Moon was inhabited. In 1776 as his observations got under way he thought he spied 'growing substances' near the Moon's Mare Humorum 'and this I now believe to be a forest'. His excitement grew: 'I am almost convinced that those numberless small Circuses we see on the Moon are the works of the Lunarians and may be called their Towns....' Having also noted to his own satisfaction metropolises, vegetation and turnpikes on the Moon, Herschel spread his net wider. He was soon to write of inhabitants of 'the Georgian planet' (ie Uranus) and a plurality of solar systems warmed and sustained by stars, of 'numberless globes that serve for the habitation of living creatures'. Herschel's pluralism was based partly on actual, if perhaps imaginatively interpreted, observation, and partly on analogy and supposition; he admitted that the plurality of solar systems 'can never be perceived by us'.[6] Herschel's achievements in conventional astronomy were entirely sound and it is upon them that his reputation rests. But he was at heart a pluralist and at least one modern scholar[7] has conjectured that pluralism was possibly the mainspring which drove his astronomical work. If so, he learned to modify this pluralism in his published works, astutely noting that even if mainstream astronomers were sympathetic to pluralism, their reputations rested on the more solid foundations of deductions made from the hard work of observation, at which he had few rivals.

Although scientists found it advisable to exercise a proper caution in their pluralist speculation, men and women of letters were freer to employ its possibilities more imaginatively. Pope's *Essay on Man* had earlier demonstrated the power of pluralist poetry. The questions it raised about the relative status and accomplishment of humanity in a putative 'great chain' had given food for thought to many, including Kant who was fond of quoting Pope's sentiment that on other worlds they might 'Shew a Newton as we shew an ape'.

20   *Imaginary lunarian life with pumpkin-dwellers, from Filippo Morghen's* Raccolta... in which *'Giovanni Wilkins' travels to the Moon (Italian, late eighteenth century).*

Pluralism, literature, and new lights

The work of the popular English literary commentator on pluralism, the Rev Edward Young, rector of Welwyn was quoted for well over a century thereafter, particularly his assertion in *Night Thoughts* (1742) that 'An undevout astronomer is mad.' Young observed a Universe which had 'one sun by day' but 'by night ten thousand shine', in which

> *...what swarms*
> *Of worlds that laugh at earth! immensely great!*
> *Immensely distant from each other's spheres!*

Upon which extraterrestrials lived:

> *Far other life you live, far other tongue*
> *You talk, far other thought, perhaps, you think...*

A plurality of worlds was an attractive theme to German poets, particularly those aware of developments in science, most notably Friedrich Klopstock. In his *Messias* ('Messiah', written 1748–73) the story of Christ is set in a pluralist Universe where, conveniently for

Christological technicalities, the inhabitants of other worlds have not experienced a Fall. The other worlds were in certain respects Earth-like in their flora and fauna, even in the outward resemblance of their inhabitants 'people like us in form, but full of innocence, not mortal people... undegenerate children' with 'hearts pounding (*klopfenden Herzen*) with desire to emulate the Father's virtue'.

The Moon was a favoured setting for pluralist literary flights (**20**). It supplied the location for opera libretti such as Haydn's *Il Mondo della Luna* ('The World of the Moon', 1777) — also the title of a poem by Saverio Bettinelli, and the farce *der Mondkaiser*, (The Emperor of the Moon, 1790) by Friederike Helene Unge. The use of the Moon as a setting for satires and other social comment, a trend already set by Bishop Godwin and Cyrano de Bergerac, bore abundant fruit in the eighteenth century in such works as William Thomson's *The Man in the Moon, or Travels to Lunar Regions* (1783) and Sommer's *Robinsons Luftreise nach dem Mond...* ('Robinson's Aerial Journey to the Moon', published in Vienna in 1785). Further afield, the newly discovered Uranus was visited in *A Journey Lately Performed... in an Aerostatic Globe... to the Newly Discovered Planet, Georgium Sidus* by 'Vivenair' (1784). In some of this literature the Earth travellers encounter aliens. Elsewhere they do not, although, in most cases, the other worlds are not only visitable but habitable and therefore the products of fashionable pluralist philosophy.

Beyond literature and natural science, pluralism started to support mystic and religious cults, the most widely spread of which was the Swedenborgian Church. The idea of other worlds being inhabited by souls or spirits, either from this Earth or native to themselves, and occasionally set in hierarchies of relative perfection, was a typical pluralist notion, although often set within wider cosmological themes. The Swedish scientist Emmanuel Swedenborg used this idea in propounding his religious faith, which he claimed had been brought to him in a series of visions by spirits from other worlds. His cosmic system consists of worlds inhabited by humans and surrounded by the souls of the departed, such as had communicated with him. The other-worldly humans differ in detail from us: Mercurians are thin; Martians are black-and-yellow. The Saturnians 'live on the fruits and pulses which their earth produces... they are lightly clothed, being girt with a coarse skin or coat, which keeps out the cold.'[8] Swedenborg sometimes based his assertions on remarkable, pedantic detail: 'supposing there were a million earths, and on every earth three hundred million men... to every man or spirit were to be allotted... three cubic ells.'

Pluralism, ever pliable and obliging to its adherents could also be used to attack religion. Tom Paine, anxious to administer a good drubbing to established authority, whether secular or religious, seized upon pluralism to discredit Christianity. He poured ridicule on Christianity within the pluralist Universe he described in his *Age of Reason* (1793). To Paine, astronomy suggested that 'each of the fixed stars is also a sun' the centre of other systems of planets. This arrangement could support his theism whilst undermining Christian notions of a unique Earth: a Fall, Incarnation and Redemption. He asked: 'are we to suppose that every world in the boundless creation had an Eve, an apple, a serpent, a redeemer?'

Paine hit the vulnerable spot which had long caused organised religion to suspect pluralism, even though pluralist apologetics had devised a range of working accommodations with Christianity, from the medieval principle of plenitude to Derham's natural theology. It is thus not surprising that religious authorities as diverse as John Wesley (1703–1791) and the Russian Orthodox Church pronounced against pluralism. Confiding to his *Journal*, Wesley, at one time a moderate pluralist, said it was a reading of Huygens that brought him to doubt life on 'secondary planets' (satellites) and eventually primary planets, 'the whole ingenious Hypothesis, of innumerable Suns and Worlds... vanished into thin air'. The Russian Orthodox Church, horrified at the pluralist assertions of the scientist Mikhail Lomonosov asked Catherine the Great to condemn pluralism: 'none shall dare to write or print anything... of the multitude of worlds...' Another source of anti-pluralism, the Jesuit F. X. de Feller (1735-1802) put his finger on the central weakness of pluralism, the absence of any clear evidence of life having been actually detected on the Moon, or of planets orbiting 'other suns.'[9] De Feller suggested that the heavens were created for Earth-bound humanity alone, by aiding navigation and stimulating pious thought. He had other, theological, objections to pluralism which harked back to earlier debates: for example that the Incarnation and the status of Earth were unique in the cosmos.

These criticisms swam, however, against a massive, popular and powerful tide. The German philosopher Johann Gottfried von Herder, steeped in pluralist thought, wondered about the relationship of humanity with 'organized beings on other worlds', about the migration of souls to bodies on different worlds, about the possible inward movement of souls to inhabited suns, and about possible future life on the Moon. He stood as it were on the cusp between the Age of Reason and the coming *Sturm und Drang* of the romantics. Pluralism was about to face its own storm and stress, during a century containing what was at once one of its golden ages, but also one of its times of greatest trial.

References

1. 'On the Nature and Construction of the Sun and Fixed Stars'*; Philosophical Transactions* of The Royal Society, 85, 1795.

2. *Lettre sur les Aveugles*, 1749.

3. There exists a complex historiography surrounding Wright's work; his reputation has been rather diminished by twentieth century scholarship. For a critique of Wright, see Stanley L. Jaki: *The Milky Way...* 1973, and M. Hoskin 'The Cosmology of Thomas Wright of Durham'; *Journal for the History of Astronomy*, I, 1970.

4. Wright had also made a similar but undeveloped suggestion, in any case unknown to Kant; it was not in the review Kant had read in 1751.

5. In his *Astronomy*, Vol II of 1764.

6. All references are from *The Scientific Papers of William Herschel*, ed J. L. E. Dreyer, 1912.

7. Michael J Crowe, *The Extraterrestrial Life Debate 1750-1900*.

8. Emanuel Swedenborg: *The Earths in Our Solar System*, 1909 edition drawn on the Latin of the *Arcana Coelestia*, 1749-1756.

9. *Observations Philosophiques sur les Systèmes de Newton et Copernic, de la Pluralité des Mondes*, 1771.

7 High pluralism, 1800-1850

Astronomers and a plurality of worlds

The high tide of the metaphysically-based 'old pluralism' with its heavy reliance on analogy, probability and religious insight came in the first half of the nineteenth century. Science was as yet only able to offer it modest support. New religious doctrines embodied pluralism into their doctrines as the Swedenborgians had already done. For the first time truly popular pluralism, a pluralism of the masses, made its appearance. These changes occurred as much in America as in Europe; the first steps towards America becoming the pre-eminent centre of world pluralism.

The previous generation of pluralist astronomers, men like William Herschel and Johann Bode, was fading from the scene but new recruits were to hand. Karl Gauss (1777–1855), the 'prince of mathematicians' and director of Göttingen Observatory wondered whether there was some kind of life after death, located on celestial bodies, and whether there was ordinary life on the Moon. In a letter of 1822 to another pluralist astronomer, Wilhelm Olbers, Gauss considered the possibility of signalling to the Moon by means of an adapted heliotrope of '100 separate mirrors'. There has long been a legend (for which there is circumstantial but soft evidence) that Gauss also proposed signalling to the Moon by means of planting a vast right-angled triangle of wheat, and squares of trees, in the hope that rational lunarians would recognise this demonstration of Pythagoras's theorem, the 47th proposition of Euclid. The director of the Vienna Observatory, J. J. von Littrow has also been credited with the plan of digging a large, oil-filled trench of geometric design in the Sahara desert, possibly a circle or square, and firing it by way of alerting Moon-dwellers to human existence and rationality.

Like Gauss, Olbers was measured in his pluralist thought: it was 'very probable that the Moon is inhabited by living, even rational creatures' and that there was vegetation there similar to that of the Earth. But popular interest is not generally fired by measured thoughts; it was the more excited speculation of Franz von Paula Gruithuisen which caught the public imagination, including that of Prince Metternich, chancellor of the Austrian Empire. Gruithuisen was a physician by training but an astronomer by inclination and later by profession, a prodigiously busy one who wrote and edited many dozens of professional papers. In 1824 he published a paper on the 'Discovery of Many Traces of Lunar Inhabitants, Especially of their Colossal Buildings' in which he laid claim to numerous observations supporting the concept of lunar life: different bands of vegetation suggesting variations in climate; paths left by lunar animals; features which might well be great walls, roads and cities. Whilst the reading public took readily to these revelations, the cautious pluralists became concerned, even horrified, that such claims if unfounded, would serve to discredit the idea of a plurality of worlds. Gauss criticised Gruithuisen's 'mad chatter'; Olbers regretted his unrestrained powers of imagination. Gruithuisen was undaunted by these reactions and pressed on to populate Venus and, giving his imagination even freer rein, suggesting that traces of major Venusian festivals could be detected by observers on Earth; massive bonfires possibly marking the coronation of a great ruler: 'another Alexander or Napoleon comes to supreme power on Venus'.

One of the most accomplished and influential astronomers of the day, Sir John Herschel (1792–1871), son of Sir William Herschel, was also a convinced pluralist. Early in his career (1833–38) he worked at the Cape of Good Hope carrying on his father's task of meticulous stellar observation, but in the Southern hemisphere. He published the first edition of what became the mid-Victorian standard work *Outlines of Astronomy*.[1] The popularity of this much-translated work carried forth pluralist sentiments to a world-wide audience of amateur and professional astronomers, as well as general readers. John Herschel vied with his father in the confident nature of his pluralism; like William Herschel he supported the idea of life on a dark, cool world within the Sun. At a time (1861) when astronomers argued about the 'willow-leaf patterns' on the surface of the Sun, Herschel added the spice of pluralist rhetoric, suggesting that these patterns might be alive: 'we cannot refuse to regard them as organisms of some peculiar and amazing kind...'.

John Herschel also favoured the idea of life on the Moon, although he was far more restrained than Gruithuisen, with one notable exception. Although he admitted that there were apparently no clouds, and possibly no atmosphere on the Moon, he seized upon the suggestion in 1856 of the mathematician and astronomer Peter Hansen that the Moon might have an off-centre of gravity so that its air and water had literally gravitated to its dark and unseen side, where therefore life might possibly reside. His writings were peppered with pluralist thoughts: the Moon might once have borne air and water on its Earth-facing side since it had level regions 'apparently of a decided alluvial character'; telescopes might one day detect traces of life upon its surface; the brightness of Mercury and Venus suggested cloud-laden atmospheres which might 'filter' the glare of the sun's rays — but for whose eyes? The pluralist implication is clear. These suggestions, in the first edition of his *Outline of Astronomy* (1849) are preceded by the thought that the wide range of physical conditions in the solar system might maintain a similarly wide range of 'animal and intellectual existence and happiness'. Even so, Herschel knew well the bounds of pluralist guesswork as compared with the deductions of observational science: 'Quitting, however, the region of mere speculation, we will now show what information the telescope affords us...'.

The scientific opinion-formers in Herschel's lifetime reinforced the pluralist tastes of the educated, and increasingly of the new part-educated public which could purchase, share or borrow books, periodicals and pamphlets. Popular poets continued to reflect pluralist ideas. Wordsworth employed conventional pluralism when he wrote of a red-haired race of Mars, or of decayed towns in Saturn thronged by 'melancholy Spectres ...'. Byron, Shelley, and Erasmus Darwin made pluralist allusions as did Coleridge who

21 *Rev Dr Thomas Chalmers whose unmatched oratory and writing, dating mainly from his Glasgow days (1815–1823), were a vital source for later Victorian pluralism.*

characteristically wavered on the central tenets of the subject: 'I could never feel any force in the arguments for a plurality of worlds...'.[2]

The Kirk and life on other worlds

Still more forceful pluralist propaganda came from the oratory and writing of two Scottish divines, principal ornaments of the long and remarkable record of Scottish pluralism. Both men, Thomas Chalmers (**21**) and Thomas Dick, knew their subject well; Chalmers was a competent and well-read mathematician; Dick a leading amateur astronomer. They were also masters of a vigorous, persuasive prose which conveyed their thoughts to immense audiences in Europe and America in repeated editions and reprintings of their works through much of the century. In addition Chalmers was the leading evangelical preacher in Scotland, well able to hold and move large congregations.

Chalmers delivered a series of lunchtime sermons to packed congregations at the Tron Church, Glasgow, published shortly afterwards as *A Series of Discourses on the Christian Revelation Viewed in Connection with the Modern Astronomy* (1817). In spite of the title and his drawing on up-to-date astronomical work, the seven sermons used assertion and analogy in typical pluralist style. He told his packed audience that planets 'must be mansions of life and intelligence' and that the far recesses of the Universe contain other worlds, lit by other suns 'and the sky which mantles them is garnished with other stars'. Chalmers' main objective was religious, not astronomical — his pluralism addressed chiefly the role and place of humanity in the Universe and particularly 'Knowledge of Man's Moral History in the Distant Places of Creation'. His argument gave an unusual theological twist to the long and unsettled dialogue between pluralism and Christianity: very possibly the Earth is the arena of a struggle between the powers of light and darkness duly witnessed by the rest of the Universe. The Crucifixion therefore gives this one 'insignificant' planet an 'importance as wide as the Universe itself'.

Although the splendour of Chalmers' oratory cannot now be recaptured, two samples from the *Series of Discourses* may hint at his powers of communication. Why should we suppose, he asks, that the 'Great Architect of Nature' calls other worlds, 'these stately mansions', into existence and leaves them unoccupied? He takes the analogy of a distant, overseas country on Earth: 'We are too far away to perceive the richness of its scenery, or to hear the sound of its population. Why not extend this principle to the still more distant parts of the universe?' Later he adds presciently: 'Who shall assign a limit to the discoveries of future ages?... The day may yet be coming when our instruments of observation shall be inconceivably more powerful. They may ascertain still more decisive points of resemblance... They may lay open to us unquestionable vestiges of art, of industry, and intelligence' — precisely that to which many astronomers were aspiring nearly two centuries later.

Chalmers' *Series of Discourses* ran to 20,000 copies in a single year, but its author moved on to other fields and had little more to add on the plurality of worlds. The Rev Thomas Dick (**22**) on the other hand produced a steady stream of popular scientific and astronomical works for some 30 years. Pluralism was explicit in many of them and became known, often for the first time, to a large reading public in Europe and America. American acolytes who managed to visit him in Dundee or Broughty Ferry included Ralph Waldo Emerson and Harriet Beecher Stowe; the Union College at Schenectady, NY awarded him an honorary doctorate. Dick's writing, like that of Chalmers, was persuasive and readable, presenting a singular blend of theology, science and religious mysticism. His first work, *The Christian Philosopher* (1817) employed natural theology, attempting to demonstrate the existence of a benevolent and omnipotent God through His works. It touched on many branches of science and technology, but gave pride of place to astronomy 'this sublime science', the study of which would reveal 'this amazing scene of divine workmanship... for there can be no question that every star we now behold... is the centre of a system of planetary worlds, where the agency of God and His unsearchable wisdom may be endlessly varied...'. It was translated into Welsh: *Yr Anianydd Christionogol* (1842) one of the first popular expressions of pluralism in that language.

Dick, a teacher, later turned to full time writing. His interests ranged widely although astronomy remained central. In such books as *Celestial Scenery* (1837) and *The Sidereal Heavens* (1840) he argued eloquently for pluralism, usually basing his claims on analogy — 'There is a general similarity among all bodies of the planetary system...' — or teleology — 'The most glorious and magnificent scenes displayed in the firmanents of the remoter

22   The 'Christian
      Philosopher' and
      high pluralist,
      Rev Dr Thomas
      Dick of Broughty
      Ferry, near
      Dundee — one of
      the most widely
      read and
      influential
      pluralists of
      Victorian times.

planets' were surely designed for observation 'by intellectual beings ... intelligent minds to whom such a display is exhibited.'

Dick's arguments may have been pluralist stock-in-trade, presented with unusual clarity; his signal contribution to pluralism was to quantify the implications of these propositions, the first pluralist to do so on any appreciable scale, unless one counts Buffon's earlier long-range estimates of the duration of planetary life-spans. In Dick's *Celestial Scenery* (**23**) he took the population density of England as the cosmic standard — an unusual choice for a Taysider — and estimated Mercury to have a population of 8.9 billions; Venus 53.5 billions; and the Moon 4.2 billions. His calculations went much further: he estimated the population of Saturn's rings and the moons of Uranus, for example. In *The Sidereal Heavens* he even attempted to estimate the population of the Universe.[3] Initially slightly cautious in his pronouncements (he cast doubts on Gruithuisen's 'discovery' of lunar cities) Dick became an audacious and extreme pluralist, later suggesting that even comets might be populated. Far from being harbingers of bad tidings he thought they might bear 'millions of happy beings to survey a new region of the Divine empire'.

Dick avoided the touchy theological controversies surrounding pluralism and the Incarnation. Instead, he preferred to speculate that some extraterrestrials might be angels, seraphs, 'intelligences' and other beings in varying states of grace. His potent cocktail of

23   Victorian space art: Saturn's rings as seen from its equator, a none-too-accurate reconstruction for Thomas Dick's Celestial Scenery *(1837), a splendid prospect 'for thousands of millions of rational inhabitants, who employ their faculties in the contemplation of the wonders which surround them, and give to their Creator the glory which is due to His name.'*

science and cosmic mysticism appealed to vast audiences. *Celestial Scenery* ran into six editions and some 25 reprintings; *The Sidereal Heavens* was similarly successful, not that the 'Christian Philosopher' benefited in financial terms; he struck bad bargains with publishers and was eventually assisted by a Civil List pension and a subscription raised by his admirers (**24**).

Thomas Dick's pluralist Christianity was only one of the connections between religion and the old pluralism during its last years. Ralph Waldo Emerson, an enthusiastic reader of Dick's works, thought that astronomy, as well as exalting 'our views of God' was a good corrective to human conceit and general anthropocentrism, particularly since the 'beings that inhabit Saturn, Jupiter, Herschel, and Mercury' were organised wholly differently from humanity. Possibly they had 'far more excellent endowments than He has granted to mankind'. None of this would have come as a great surprise to many American audiences. They were already avid readers of Dick's works, and also of their fellow-American the Rev Timothy Dwight of Yale. His posthumously published sermons *Theology Explained and Defended* (1818) contained numerous supportive references to pluralism couched in prose which rivalled that of Chalmers: 'With His own hand he lighted up at once innumerable suns, and rolled around them numerous worlds... Throughout His vast empire, he surrounded His throne with Intelligent creatures...' Dwight's seventeenth sermon presents one of the clearest and most fully-stated cases of relating natural theology to pluralism. The stars, he points out: 'give light, and motion, and life, and comfort to systems of worlds...'. He grappled with the problem of the Incarnation by suggesting that

24   Mills Observatory, Dundee; one of Britain's few public observatories, financed by the Tayside capitalist John Mills, an admirer of Thomas Dick and his pluralist arguments and a fellow member of the Secession Kirk. Another admirer was Ralph Waldo Emerson who thought like Mills: 'I hope the time will come when there is a telescope in every street.'
(Courtesy: Mills Observatory, Dundee)

on Earth alone there exists 'a singular and astonishing system of Providence; a system of mediation between God and His revolted creatures' — echoes of Chalmers' philosophy, but with subtle differences. Like Chalmers and Dick, Dwight was a popular author: his *Theology Explained...* was still being published in 1924.

Pluralism and new faiths

During this phase of high pluralism the theology of some less conventional branches of Christianity was illuminated by pluralist light. The Swedenborgians, by now the New Jerusalem Church, restated their case in Samuel Noble's *Astronomical Doctrine of a Plurality of Worlds* (1828) — in part a counterblast to Tom Paine's anti-Christian pluralism and the anti-pluralism of John Hutchinson (1674–1737) and his followers, whose interpretation of the Bible caused them to reject Newton's work and to doubt pluralism. Swedenborg societies grew in number; there were 54 of them in America by 1850.

Another church which employed pluralist arguments, but which grew to a greater size, was the Church of Jesus Christ of the Latter-day Saints, the Mormon Church, founded in 1830. Although *The Book of Mormon* itself contains no pluralist assertions, later fundamental texts of the church did so: *The Doctrine and Covenants*, and *The Pearl of Great Price*. The founder of the Mormon Church and its 'First Prophet', Joseph Smith, claimed to have a series of visionary revelations including one which told him that angels 'reside in the presence of God, on a globe like a sea of glass and fire...'. This and similar passages have given rise to speculation that Joseph Smith may have been influenced by the more mystical writing of Thomas Dick, notably his *Philosophy of a Future State* with its theory of a great central body governing the machinery of the cosmos. Even if such a connection were proved it would not necessarily discredit Smith or any other of Dick's readers; theology is famously subtle and apologists could always argue that the ways of inscrutable providence are hid from us by what Chalmers suggested was 'the darkness of our minds'.

One religious community which drew direct inspiration from pluralist arguments was the Seventh-day Adventist Church. The publication *The Opening Heavens* (1846) by one of its founders, Joseph Bates (1792–1872), shows that he was well-acquainted with the pluralism of Thomas Dick and others and used it to support his claim that the second coming of Christ would be through the great nebula in Orion. The 'prophetess' of the church, Ellen White *née* Harmon had pluralist visions; in 1849 she stated:

> The Lord has given me a view of other worlds... an angel attended me from the city to a place that was bright and glorious. The grass of the place was living green, and the birds there warbled a sweet song. The inhabitants... were of all sizes; they were noble, majestic, and lovely...

The extent to which mystical pluralism had spread in these years is vividly illustrated by the case of the eminent scientist Sir Humphry Davy, president of The Royal Society. Towards the end of his life (he died in 1829) he wrote *Consolations in Travel, or the Last Days of a Philosopher* in which the writer is taken on an exploration of the cosmos where he finds souls transmigrating between planets. He described the Saturnians in some detail, creatures not unlike sea-horses with 'strangely convoluted elephantine proboscices', but nevertheless enjoying intellectual superiority over humans. Davy was not writing science fiction: he insisted that his book contained 'certain truths that cannot be recovered if they are lost'.

Science and Logic

On the other hand, a pluralist effusion from the Danish scientist Hans Christian Oersted *The Soul in Nature* (published in 1852, but embodying the author's earlier pluralist thoughts) based its case not in dreams, but in physics and chemistry. Given the ubiquity of matter in the Universe, Oersted argued, and given that it has resulted in life on Earth, life will appear elsewhere given similar favourable conditions. Extraterrestrial beings might not resemble humans physically, but since everywhere 'reason should develop into self-consciousness', extraterrestrial science, art and morality should resemble ours. There was a strong religious element in Oersted's pluralist writing although he attempted to keep his science separate from his religious and metaphysical ideas. Nevertheless they inevitably coalesced in his pluralist Universe evolving under the direction of an 'Eternally Creating Spirit'.

Although pluralism was thus increasingly well-supported by science, philosophy, the arts, and a wide range of religious faiths in Europe and America, it still had its critics and problems. Two examples from different ends of the cultural spectrum may illustrate this. In 1843, John Stuart Mill published his influential *System of Logic*. In tackling the problem of analogy (Book III, Chapter 20) he chose as his chief example the doctrine of a plurality of worlds. Mill argued that analogy, much favoured by pluralists, is not sufficient proof — although a 'balance of probabilities' may be of assistance in coming to a conclusion. If the Moon, for example, lacks the 'indispensable conditions of animal life' prevailing on Earth then a conclusion that there is, indeed, life on the Moon 'must be an effect of causes totally different' from those on Earth. He continued by stating that we know so little of the 'immeasurably multitudinous' conditions prevailing on the planets of the solar system that we 'cannot attach more than trifling weight' to the few resemblances we can actually observe. This led him to the famous conclusion that whilst analogy may initiate or guide our discoveries, it cannot class as reasoned proof: it had 'the highest philosophical value' as 'a mere guide-post, pointing out the direction in which more rigorous investigations should proceed'. The Prussian astronomer F.W. Bessel had already warned (1834) against the pluralists' frequently excessive and unwarranted claims that each and every planet was populated; a 'fruitless attempt [which] blackens with fantasies a science which is so rich in attainable facts'.

The Great Moon Hoax

Shortly afterwards, and six years before Mill drew attention to the flaws in one of pluralism's longest-established props, an altogether different publication threw doubts on pluralism by demonstrating its disturbing capacity for what later critics would term the 'hype factor'.

In New York, 1836, the proprietor of the failing daily newspaper the *New York Sun*, Benjamin H. Day, was casting about for an idea to increase circulation. He promoted an English immigrant, Richard Adams Locke, and charged him with the salvation of the paper. He had made an inspired choice. Locke pulled off one of the great coups of journalistic history. Like many innovators he combined creatively elements that already existed. These included: a strong and growing popular interest in pluralism; the attested fact that Sir John Herschel was at work in the Cape of Good Hope surveying the Southern skies; and that people are generally too busy to check every detail of obscure or inaccessible sources. He quoted a fictitious 'Dr Grant' of the *Edinburgh Journal of Science* — which had ceased publication. The *Edinburgh Courant* which ostensibly passed on Dr Grant's revelations did exist, but was hard to obtain in New York. These 'sources' revealed that Herschel had actually discovered life on the Moon: dark red flowers, yew-like trees, and quadrupeds resembling bison, or yaks. Locke was a thorough market-researcher and, by mingling with the crowds swarming outside the offices of the *Sun* he soon sensed what they really wanted to know. With sureness of touch he rationed the news in subsequent editions: Herschel had seen slender obelisks made of crystal; then he detected intelligent bipeds, a type of hybrid between a beaver and a human; then finally actual humanoids, like *homo sapiens* but with bat-wings, *vespertilio-homo* (**25**). By this time the *Sun* had the largest circulation in the world, 19,360 copies daily, 2,300 copies ahead of its nearest rival, *The Times* of London.

25   *Lunarians: bat-people of the Moon; an illustration inspired by Locke's Great Moon Hoax of 1836.*

The 'Great Lunar Hoax' was exposed shortly afterwards — Locke confessed all to a fellow-journalist in a New York bar, possibly the worse for wear according to some accounts. Even so, the lunar hoax was reprinted in book form and widely translated into French, German, Danish, Italian and also into Welsh: *Hanes y Lleuad; yn Goson Allan y Rhefeddodau a Ddarganfyddwd gan Syr John Herschel* (Llanwrst, 1836), narrowly antedating the Welsh translation of Dick's more general work. The lunar hoax depended on a large and curious public with sufficient purchasing power and literacy. It demonstrated the international spread of the pluralist paradigm and the hopes, wonders and speculation it generated among the increasingly literate masses. It also relied upon technology: Locke's mumbo-jumbo about Herschel's gas-powered telescope, and the *Sun's* newly-installed rotary steam-presses which could meet the insatiable demand for more Moon-material each day 25–29 August 1836.

Mill and Locke had, in their individual ways, sounded warning notes about the claims and background assumptions of the old pluralism. At this moment of its greatest strength it broke against a series of formidable obstacles.

References

1   *Treatise on Astronomy*, part 43 of the *Cabinet Cyclopaedia*, 1833; amplified into *Outlines of Astronomy*, 1849, much translated — into Arabic and Chinese, *inter alia*.
2   *Table Talk*, 1834.
3   Apparently 60,573,000,000,000,000,000,000 beings.

8 Old pluralism into new, 1850-1880

Half-way through the nineteenth century traditional pluralism was overtaken by a new form which, although embodying many of its predecessor's methods, was different in one crucial respect: it was increasingly based on science rather than metaphysics or religion. This came about for two reasons: first, pluralism faced its first major intellectual challenge since the Middle Ages when William Whewell, an eminent Cambridge scientist and divine, published a severe critique of the usual arguments for a plurality of worlds. Secondly, fresh developments in science informed the pluralist debate. From that time mainstream pluralism has been to a greater or lesser extent a predominantly scientific issue.

Other developments were also changing the intellectual and social foundations of pluralism. The new-order world of 'carboniferous capitalism' was able to educate and inform ever larger numbers of people, as it financed and equipped the laboratories and observatories which generated data for the new pluralist debates. Improvements in optical glass had led to larger telescopes; by 1838 the Imperial Russian observatory at Pulkovo had installed a 38cm refractor. Shortly afterwards the citizens of Boston raised sufficient funds to purchase a clone, duly presented to Harvard University in 1847. This was also the great age of the 'grand amateur' astronomers, beneficiaries of capitalism who invested in fine instruments and could often do important, pioneering work with them: Nasmyth and Lassell in Britain, for example. In 1830 some 24 private observatories were listed in Britain; by 1866 this had grown to 48.[1]

Most professional astronomers were concerned with ever more precise celestial measurements, for example the distance to the stars first announced by Bessel in 1838, and the discovery of the planet Neptune by mathematical reasoning in 1846. Even so, there were indications of emergent astrophysics; John Draper, like the raffish R.A. Locke an English immigrant to America, first photographed the Moon in 1840 and recorded the solar spectrum in 1841. In Ireland, the Earl of Rosse constructed a giant reflector telescope at Birr Castle, Parsonstown — its size was unrivalled for some seventy years — and discovered that distant nebulae were sometimes of a spiral shape, resolvable into myriads of stars.

Whewell's bombshell

It was against this background of improving science, and within a paradigm of high pluralism that William Whewell threw his bombshell 'into an army resting on its victorious march' as the American, Rev Theodor Appel later put it. Whewell, Master of Trinity College, Cambridge a leading scientist and respected moral philosopher, had hitherto been sympathetic to pluralism. Influenced by Chalmers he had written a popular treatise in 1833, a closely-argued commentary on the relationship of astronomy to natural theology in which pluralist considerations, although lightly treated, were nevertheless present. He thought that other planets 'may... be the seats of vegetable and animal and rational life... so many provinces in the same empire'.

The publication of Whewell's *Of the Plurality of Worlds: An Essay* in 1853, however, struck the very heart of pluralism. It has sometimes been hailed as the first scientific critique of pluralism, well-informed by contemporary astronomical findings and sharp in its criticism of pluralist methodology. Nevertheless it was strongly suffused with theological arguments. Recent work[2] suggests that Whewell's point of departure was essentially religious; he employed science and logic to support the proposition that God had created but one home in the cosmos for intelligent life: planet Earth. If this were so, Whewell and his widely-read book (which ran to five British and two American editions) were transitional rather than representative of a clean break with pluralism's metaphysical past.

Whatever Whewell's underlying motives, the furore he created was considerable; some twenty responding books, including at least one best-seller, and up to fifty articles and long reviews in journals commented on his argument. Nearly three-quarters of this mass of literature was opposed to him. But what had he said to generate such excitement? His chief assertion, couched in vigorous and compelling prose, was that the uniqueness of humanity in the 'growing magnitude of the known Universe' was a mark of God's special care, of humanity's special place in the divinely-ordained scheme.

Whewell had no difficulty in contemplating life on other worlds; possibly Jupiter might be a sphere of water inhabited by 'boneless, watery, pulpy creatures' and Mars might be like the early Earth, the home of dinosaurs. But to him Earth alone bore sentient, intelligent life able to receive an incarnate God, 'the arrival of this especial message... forms the great event in the history of the Earth'. Claims that there were similar beings to humans elsewhere in the Universe were dismissed as 'an act of invention... as coherent as a fairy tale... purely imaginary and arbitrary'. In spite of the discovery by astronomers of the vast size of the Universe, the Earth 'can not, in the eyes of any one who accepts this Christian faith, be regarded as being on a level with any other domiciles'.

Whewell cited science in support of his thesis: geology demonstrates the immense tracts of time passed before humanity appeared on Earth, but a lavish endowment of resources for given ends is part of the divine plan. Whewell selected widely from available astronomical data, but interpreted them to support a case already stated. He was able to do this partly because astronomers, working at the cutting edge of optical technology, understood imperfectly the nature of the great nebulae they were observing. Lord Rosse's work, quoted by Whewell, suggested that nebulae were composed of many stars, although Sir John Herschel working at the Cape had noticed that the Magellanic Clouds seemed to be composed of some stars but also many patches of 'nebulous matter'. Whewell averred that Rosse's findings did not necessarily suggest nebulae composed of myriads of 'self-luminous' stars but possibly of mere 'lumps of light' or 'dots' incapable of holding planetary systems. Noting the discovery of variable stars, that is of those whose light radiation waxes and wanes, Whewell suggested, incorrectly, that their character resulted from their repeatedly turning a 'dark side' to the Earth, hence they were quite different from the Sun in their construction.

Whewell was anxious to demonstrate that the Universe was not a particularly congenial home for intelligent life, and in any case 'The majesty of God does not reside in planets or stars ... which are ... only stone or vapour, materials and means.' He listed actual and potential human achievements including 'the thought of Rights and Obligations ... Duty and Virtue, Law and Liberty... the elevation of our fellow-Citizens' which were attributable to 'a body of beings more worthy of account than millions of mollusks... lizards and fishes, sloths and pachyderms, diffused through myriads of worlds'.

Although *Of the Plurality of Worlds* was at first an anonymous publication, its authorship was soon known. The prestige of its writer, the cogency of its arguments, and the effective deployment of religious and scientific material, engaging to the mid-Victorian mind, made Whewell's book the greatest *cause célèbre* of pluralist theory since the times of Bruno and Galileo.

The immediate response, mainly in Britain, to Whewell's 'Essay' (it was 279 pages long) represented a good cross-section of the pluralist debates and positions at the time. The *British Controversialist*, a finely named journal in the roaring days of British individualism, debated Whewell's thesis throughout 1855 with exchanges from traditional pluralists like 'Philalethes' who employed the analogic arguments about the abundance of life on Earth probably being replicated elsewhere; from Whewellites like 'S.S.' who drew attention to the weaknesses of analogic argument; and the compromising 'Threlkeld' who used analogy to support the notion of life on other planets, but neither on the Sun nor the Moon.

The Rev Robert Knight of Polesworth set out to rescue Christian pluralism in his *The Plurality of Worlds: The Positive Argument from Scripture* (anon, 1855) a work influenced by religious mysticism. Knight thought that Whewell's arguments tended, like those of Tom Paine, to undermine the Christian faith, the tenets of which might be transported to other worlds by means of angels acting as messengers. If astronomers had not yet found extraterrestrial life they might do so one day; in any case the spiritual essence of extraterrestrial beings might reside in beings not 'palpable' to human sight.

Scientists also had much to say for and against Whewell. Sir John Herschel, a personal friend, was not unduly put out, or at least he endeavoured to give that impression. Having stated in a letter to Whewell: 'I should not have thought there was so much to be said on the non-plurality side of the question'. He went on to speculate about the possible denizens of Whewell's aquatic Jupiter: 'to what perfection they may have brought the science of hydropathy'. Herschel's response seems to have been one of mild bewilderment offset by friendly loyalty; he never reviewed *Of the Plurality of Worlds*.

The American Denison Olmsted was sympathetic to Whewell's argument, commenting that 'the telescope...has added nothing to the amount of evidence in favor of

E. R. PAIGE, SCIENTIST.

CHICAGO:
Central Music Hall. CIRCULAR OF THE SLAYTON LYCEUM BUREAU. SEASON of 1880-1881.

WILL DELIVER HIS LECTURES:

ORIGIN AND GROWTH OF WORLDS.

ORIGIN AND DEVELOPMENT OF LIFE.

OTHER WORLDS & THEIR INHABITANTS.

All Brilliantly Illustrated

BY OVER 60 DIAGRAMS AND PAINTINGS OF THE SOLAR SYSTEM,
ROCK SECTIONS AND DISTINCT ANIMAL FORMS.

26   Popular pluralism: advertisement for a pluralist lecture with visual aids, Chicago, 1880.

the doctrine that the planets are inhabited...'. Nevertheless he stayed loyal to his initial pluralism; he had been a student of Timothy Dwight. He thought that the other planets 'are also abodes of life and intelligence'. Maria Mitchell was unimpressed by Whewell and his arguments: 'There is nothing from which to reason. The planets may or may not be inhabited.' Two leading geologists, Rev Adam Sedgwick and Sir Roderick Murchison differed; the former was 'amused by it, but not convinced' whereas the latter admired the author's 'ingenious and well-reasoned reviews'.

One respondent changed his position on pluralism, or at least had earlier doubts confirmed — the Scottish minister Rev George Gilfillan. Once moved and convinced by Chalmers' pluralism, Gilfillan had come to see it as 'only a rainbow, beautiful, evanescent, unreal'. Whewell's book he found, on the other hand, full of 'manly energy, clear precision, and philosophic calm... a blow in the face of Natureworship'. But traces of his earlier persuasion led him to wonder whether the Earth might be but the initial world to be populated in a 'great colonization'.

An Edinburgh counterblast

It was, however, from another member of the notable Scottish school of pluralism that Whewell received his greatest counter-barrage. The Edinburgh physicist Sir David Brewster, who had previously sparred with Whewell, reacted fiercely. His religious pluralism had been challenged at a difficult time — his wife had died in 1850 and his faith was in crisis. Hence such remarks (re. Whewell's chapter on the fixed stars) as 'only some morbid condition of mental powers which... delights in doing violence to sentiments deeply cherished, and to opinions universally believed...'. To a considerable extent much of the post-Whewell debate was a commentary on the differences between Whewell and Brewster whose *More Worlds than One: The Creed of a Philosopher and the Hope of the Christian* (1854) outsold *Of the Plurality of Worlds* running to three editions and over one dozen printings in Britain, and three in the USA.

Brewster tended to deploy time-worn pluralist arguments but in doing so produced a good state-of-the-art compendium of high pluralism at flood tide, just as it broke. He quoted biblical references which ostensibly supported a plurality of worlds.[3] He noted that in neither the Old nor New Testaments 'is there a single expression incompatible with the great truth, that there are other worlds than our own which are seats of life and intelligence', the argument advanced previously by Campanella and Wilkins.

To Brewster, Whewell's assertion, that geology demonstrated the exuberant abundance of Nature working through long periods of time in order to produce intelligent life, was 'too ridiculous even for a writer of romance'. Whewell's interpretation of the nebulae, admittedly awry but nevertheless plausible given the optics of his day, was 'the grapeshot of assertion, banter, and ridicule'.

In offering his own case for pluralism Brewster played old hands, even overplayed them. His suggestion that geology might one day discover traces of pre-Adamite humanity was ingenious but unconvincing. He used analogy and probability, as well as teleological religious arguments which were already well-known. In asserting that there was life on the Sun, that 'every single star' was the centre of a planetary system, or that Jupiter might contain beings compared to whom Newton's intelligence would be of the 'lowest degree' he carried pluralism to heights it had rarely reached. In doing so he appealed to a wide-reading public, but at the same time lost the more measured support which was otherwise sympathetic to his general line.

One such source was the Rev Baden Powell, Savilian Professor of Geometry at Oxford, father of the founder of the world scout movement. Baden Powell was a moderate, open-minded pluralist, and later a supporter of Darwin. In his *Essays on the Spirit of Inductive Philosophy* (1855) he rejected charges that Whewell was in any way critical of religion, but cast some doubt on Whewell's use of geology in asserting the uniqueness of humanity. In trying to 'hold the balance between the two disputants' Baden Powell treated pluralism as essentially a scientific question unavoidably bound up with the use of analogy and probability, which needed to be kept in check. Exceeding his own careful rules he thought that all stars were probably the centres of planetary systems. On the other hand, his suggestion that some planets are ready for life whilst on others 'only a comparatively small advance may have been made' was clearly compatible with evolutionary theory, and one of the clearest precursors of the 'new pluralism'.

Pluralism and 'The march of science'

Although the Whewell crisis cast its shadow over scientific debate in the late 1850s, it was soon eclipsed by the far greater furore occasioned by the publication of Charles Darwin's *Origin of Species* in 1859. Thenceforth evolution replaced pluralism as the main arena for arguments about the relationship between science and religion; it also replaced pluralism as the main topic of popular scientific debate (**26**). Darwin's evolutionary theory cut both ways with pluralism. To some commentators it suggested that if life had evolved on one planet, then why not on others? This line of reasoning was to colour the imminent controversy about life on Mars. To others, including Darwin's fellow-pioneer of evolutionary theory, Alfred Russel Wallace, intelligent life on Earth was so remote an outside chance, arising from random mutations and conditions that it was probably not replicated anywhere else.

The main contribution of Darwinism to the pluralist debate was to undermine natural theology, and especially its teleological tendencies. Evolutionism suggested a cosmos in which there was (as Whewell had dared to suggest) great wastage, and in which an overall strategy was harder to detect than ever. Darwin himself was a moderate pluralist although he inclined to the view that 'the Creator of countless Universes' worked through general laws rather than by attending individually to each and every detail of creation.

Other developments in science intensified and extended pluralist debates. The spectroscope enabled astronomers from the 1860s to determine the chemical composition of the stars. Foremost amongst the early celestial spectroscopists, Sir William Huggins (**27**, **66**) encouraged pluralists when he reported in 1864 that his study of stellar spectra led him to conclude that there was 'some proof that a similar unity of operation extends throughout the universe' and that the presence of 'terrestrial elements in the stars' was probably to be found 'in the planets genetically connected with them'.

Huggins was a pluralist at the time, although he later moderated his views. Like Darwin, his work both helped and hindered pluralism: further spectroscopic analysis suggested that some of the nebulae at least were glowing clouds of gas, not 'island universes' composed of many stars. The pioneer American astronomical spectroscopist, John Draper, the photographer of the Moon, remained, however, an unambiguous pluralist to whom each star was 'a life-giving sun to multitudes of opaque, and therefore invisible worlds...'.

Another scientific revelation affected pluralism in the immediate post-Whewell years: the application of the kinetic theory of gases to the atmospheres of the planets, largely the work of the Irish physicist G.J. Stoney. Stoney suggested that gases, therefore the atmospheres of planets, escaped into space at rates determined by two factors: the molecular weight of a gas, and the gravity of a planet or satellite. Thus the relatively lightweight Moon would lose its atmosphere more quickly than the heavier Earth, a suggestion borne out by observation. Similarly, lighter hydrogen had escaped from the

27   *Sir William Huggins with his 15-inch refractor, spectroscope attached, at Tulse Hill observatory. The analysis of the chemical composition of the stars by Huggins and his wife stimulated pluralism in the late nineteenth century by demonstrating the universality of familiar elements.*

Earth's atmosphere rather than heavier oxygen and nitrogen. Stoney thought that, for related reasons, 'it is probable that no water can remain on Mars'. This was not good news for pluralists, although they were to use the concept of a Martian water-shortage later, in imaginative ways.

Whilst these tides of argument ebbed and flowed, the routine work of Fr Angelo Secchi, SJ continued at the Jesuits' Observatory in Rome. Secchi was a prominent astronomer, one of the quiet figures of history whose self-effacement conceals great influence and authority. He was a personal friend of Pope Pius IX who greatly respected his judgement and astronomical ability. He was also a keen pluralist. Almost certainly these two factors kept pluralism safe from official Catholic disfavour. In spite of its association in the eyes of some critics with materialism — and other tendencies objectionable to Catholic dogma — pluralism never appeared on the Syllabus of Errors. Catholics were thus able to enter the new, scientific, pluralist debates which many of them did, particularly in France.

Secchi was also a pioneer spectroscopist; his classification of stars according to spectral type was the first such taxonomy. Secchi's other contribution to pluralism was probably unintentional. When observing Mars in 1859 he saw two thin lines on its surface which he referred to as *canali*, channels. A fateful concept had been launched on its way; it was to have incalculable results.

References

1 *Edinburgh Encyclopaedia and Astronomical Register* of those years, respectively; noted in Allan Chapman: *The Victorian Amateur Astronomer*, 1998.

2 See for example, a full and detailed analysis in Michael J. Crowe: *The Extraterrestrial Life Debate 1750-1900.*

3 Such as *Amos* IX.6: It is he that buildeth his stories in the heaven...; *Ephesians* I. 10: That in the dispensation of the fullness of times he might gather together in one all things in Christ, both which are in heaven, and which are on earth.

9 To the canals of Mars — and after, 1880-1920

Popular Pluralism in the *Belle Epoque*

It was nearly fifty years before science-based pluralism enjoyed its first harvest, a bitter one as it turned out. During this period there was a brief, renewed excitement about life on the Moon, and an equally brief flurry of papers about whether or not traces of organically-based life had been found in some meteorites that had landed in France. Even though mainstream pluralism was now more closely informed by science, it continued to demonstrate an obliging affinity with a wide range of cultural forms: materialism, spiritualism, and new genres in literature. Popular pluralism was well-served by two literary figures who joined such writers as Fontenelle, Derham, Chalmers and Dick in its pantheon: Camille Flammarion in France, and Richard Proctor in Britain and the USA. Both were virtually household names in their time, *c*1870-1910, and both made a lucrative living writing for an increasingly literate public. It was against this background that the Martian canals controversy was to burst, after which nothing was to be quite the same.

Pluralism, spiritualism and materialism

The pluralist debates in France were fairly typical of those elsewhere, particularly regarding pluralism and spiritualism. The notion of souls inhabiting planets elsewhere in the cosmos, or of transmigrating between them was of course not new. Earlier pluralists like Thomas Wright of Durham and Emanuel Swedenborg had written copiously on the subject. The Swiss scientist, Charles Bonnet (1720–93) had put it: 'you are called one day to take your place amongst the celestial hierarchies, you will soar... from planet to planet; you will go from perfection to perfection...'. Similar sentiments became popular perhaps because of the emerging scientific tone of the new pluralism. Victor Hugo was amongst the devotees of Flammarion's popular synthesis between pluralism and spiritualism. Flammarion's argument had been inspired by Jean Reynaud who advocated a type of interplanetary metempsychosis in his *Terre et Ciel* (1854). Louis Figuier's *Le Lendemain de la Mort ou la Vie Future Selon la Science* (1871) imagined another cosmos in which the dead, having been clothed in the 'ether' which suffused space, migrated to the Sun, there to join others in shining forth the light and warmth we receive on Earth.

28 *The arch-pluralist in his prime: Camille Flammarion in 1898 from A. von Schweiger-Lerchenfeld's* Atlas der Himmelskünde.

The many writers in France and elsewhere who tried to reconcile pluralism with Christianity were increasingly aware of the challenges growing not only from spiritualism and mysticism but, more threateningly it seemed at the time, from materialism. Karl Marx's chief theoretical collaborator, Frederick Engels, was an enthusiast for science and entirely sympathetic towards pluralism. His *Dialectics of Nature* (written in the 1870s but not published for 50 years) stated 'an eternally repeated succession of worlds in infinite time is the only logical component of... innumerable worlds in infinite space'. In his *Socialism: Utopian and Scientific* he wrote of a succession of people being born and perishing wherever there were favourable conditions on 'celestial bodies'.

For these reasons amongst others, pluralism was to be tolerated, even occasionally favoured, years later in the Soviet Union — as a proposition congruent with the dialectical materialism underlying the official dogmas of Marxism-Leninism. Another German scientific materialist, David Strauss (1808-1894) presented a pluralism linked to uncompromising materialism in his *Der Alte und der Neue Glaube* (1872) which offered an evolutionary view of the cosmos containing 'not only completed worlds, but also such as are only in process of formation'.

Some theologians, having already sensed the dangers of a pluralism which could support this kind of aggressive materialism, renewed the religious case against a plurality of worlds, although they were in a minority. In France, the Abbé Joseph Filachou had rejected pluralism (1861) because it lacked 'positive proof' and was 'incompatible with all that [is]... most rational in metaphysics, aesthetics and physics'[1]. Monseigneur de Montignez, although moved by Flammarion's sincerity and eloquence, added (1865) that his pluralist spiritualism had little chance 'in the eyes of science or at the tribunal of reason'. He nevertheless supported pluralism, offering a complex scheme whereby 'the elect of the Earth' are to govern extraterrestrials because they alone have been redeemed by the Incarnation — an echo of earlier astro-theological debates.

Camille Flammarion (1842-1925), the astronomer whose persuasive eloquence influenced this debate wrote many much-translated works (**28**). A generation of Edwardian amateur astronomers in Britain cut its teeth on his best-selling *Astronomy for Amateurs* (its title significantly translated from the French, *Astronomy for Women*). Flammarion's chief pluralist works were: *La Pluralité des Mondes Habités* (1862) and *Les Mondes Imaginaires et les Mondes Réels* (1865); the former ran to 33 editions and was in print as late as 1921; the latter had 14 editions and was still in print in 1925.

Flammarion wrote florid prose which, when combined with the intrinsic interest of his subject-matter, made him one of the most widely-read pluralists of all time. His pluralism was not linked invariably to spiritualist musing; most of it was straightforward and secular, usually resting its case on up-to-date science, or at least Flammarion's singular interpretation of it. He ranged from cautiously speculative rhetoric (of lunar life):

> we cannot affirm that... there are not some changes which can be due to the vegetable kingdom... or, who knows? — to some living formations which are neither vegetable nor animal,

to flamboyant guesswork devoid of astronomical warrant:

> [of Jupiter] a charming place — continual Spring reigns on its surface. If it is ornamented with flowers which we do not doubt... they live much longer than ours...;

> [or of Mars] on which doubtless a human race now resides... These unknown brothers are not spirits without bodies... but active beings, thinking and reasoning as we do here. They... have raised cities, and conquered the arts.

and pure prose-oratory:

> and when in the East the sublime nights light up their diamond constellations... through the immensity of the Worlds, among the stellar skies... beyond the unknown regions where eternal splendour spreads... let us salute, my brothers, the sister-Humanities passing by.

Flammarion's flights of imagination were coloured by a close consideration of the physical conditions which he believed obtained on other worlds; a Darwinist of his day, he supposed that extraterrestrials adapted to particular environmental conditions and evolved accordingly. He supposed, moreover, that the cosmic evolutionary hierarchy did not necessarily end with 'inferior humanity': 'life develops without end in space'. This blend of hard science and evolutionism, lightly spiced with spiritualism and occasional passages of religious scepticism, all couched in fruity wordage, proved irresistible to his readers, particularly the anti-clerical intelligentsia of the Third Republic.

A more sober appraisal of the pluralist scene, equally readable and well-informed, was provided by the popular works of the English pluralist Richard Proctor. Proctor moved to America, settling in Florida. His astronomical readership was then almost certainly the largest in the English-speaking world. Most of his output was concerned with astronomy but he frequently inserted pluralist passages or essays, often of considerable length. Proctor was that comparative rarity, a pluralist who changed position in the face of new evidence; nearer to the spirit of good science than the more assertive, creative Flammarion.

In his *Other Worlds Than Ours* (1870, seven later editions and 29 printings) Proctor rejected as 'too bizarre' the idea of life on the Sun, but thought it 'all but certain' that the Moon was inhabited, and that all the evidence pointed to 'Venus as the abode of living creatures not unlike the inhabitants of Earth'.

At this time he shared Brewster's view that the Earth was well cared for and 'nobly planned' but within a lustrum, in *Our Place Amongst Infinities and Science Byways* (1875) he was shifting to Whewell's position, criticising pluralist arguments based on mere analogy, and adopting an evolutionist standpoint. He conceded that there might one day in the future be life on the Sun, that Mars had possibly lost any life it once nourished, and that Jupiter was en route to becoming a home for intelligent life. Proctor's evolutionary perspective was a British compromise: 'both the Brewsterian and Whewellite theories of life on other worlds gave place in my mind to a theory intermediate to them, in another sense opposed to both'.

Two lesser-known but remarkable works of the 1880s, both influenced by Proctor were *The Heavenly Bodies, Their Nature and Habitability* (1883) by the Edinburgh solicitor William Miller, and *Die Sternewelten und ihre Bewohner* ('The Star Worlds and their Inhabitants', 1884–85) by the German Catholic priest, Fr Joseph Pohle. Both writers took the then unusual step of offering a short history of pluralism, a significant development which marked the emergence of the subject as an established cultural form with its own evolutionary story. But whereas Miller employed his considerable erudition to reject pluralism, largely on Whewellite lines, Pohle supported it. Like Proctor, Baden Powell and Engels he had an evolutionary perspective: 'Each central body has its definite period of life which is enclosed... between long epochs of death.'

By the waters of Tharsis

Pluralist theories, whether these were metaphysical, religious, or scientific in their sources and reasoning still suffered a major problem: no one had yet produced verifiable evidence of life, intelligent or otherwise, existing on 'other worlds than ours'. But this was about to change.

The first straw in the wind came in a report in 1877 from the Brera Observatory, Milan. The Italian astronomer Giovanni Schiaparelli reported that he had apparently observed a network of *canali* (channels) on the surface of the planet Mars, the term first employed in

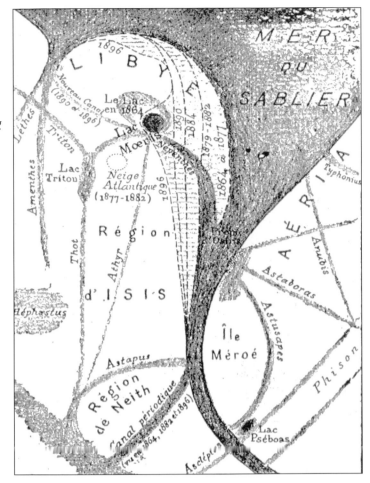

29   *Movements in the coastline of the Martian 'Mer du Sabilier' (Syrtis Major) 1864-96, and local canal networks, according to Camille Flammarion.*

this context by Fr Angelo Secchi in 1859. *Canale* was soon translated into English as 'canal', implying artificial design and construction, and it quickly gave rise to speculation and controversy.

Schiaparelli had been aided in his work by the complexities of Mars' orbit which had brought the planet nearer than usual to Earth in 1877[2] so that it presented a relatively large telescopic image even to Brera's modest 8-inch refractor. Similar proximities were to occur in 1892 and 1909, in each year re-igniting the interest and excitements of the 'discovery' of 1877. The Martian 'canal era' was thus to last for some thirty years.

Astronomers had already observed features on Mars that might be continents or seas: Schiaparelli not only confirmed these findings but ascribed to them finely sonorous classical names: Hellas, Tharsis, Mare Erythraeum, Promethei Sinus, bringing some respectable, terrestrial nomenclature to the infant science of 'areography', the Martian answer to the 'geography' of planet Earth. Schiaparelli was cautious in his public statements but it has been suggested[3] that he was possibly a closet pluralist. He may even have even intended to exploit the popularity of pluralism in his search for funding to upgrade the astronomical instruments at Brera. If so, he was neither the first nor last

*30   Percival Lowell as 'The Watchman of Mars' with his 24-inch refractor at Flagstaff,
spectroscope attached; Harmsworth's* Popular Science, *edited by Arthur Mee (1912).*

scientist to play his cards this way; if devious then certainly also rational, given the
eccentricities of public and political priority-setting. Either way he was successful; Brera
installed an 18-inch telescope in 1886. Ultimately the saga that was about to unfold
involved some of the latest instruments in the world, some with apertures of 15, 24, and
30 inches — large for the time. This working at the very edge of optical technology created
uncertainties and played a part in generating the disagreements that were to come.

Meanwhile in 1882, Schiaparelli had reported his discovery of twin *canali*, a
configuration which he named 'gemination'; he described sixty *canali* and twenty
geminations. Some of his findings were confirmed by Perrotin and Thollon at Nice
Observatory in 1886, although two years later Perrotin confused matters by announcing
that the Martian continent of 'Libya' observed by Schiaparelli in 1886 'no longer exists

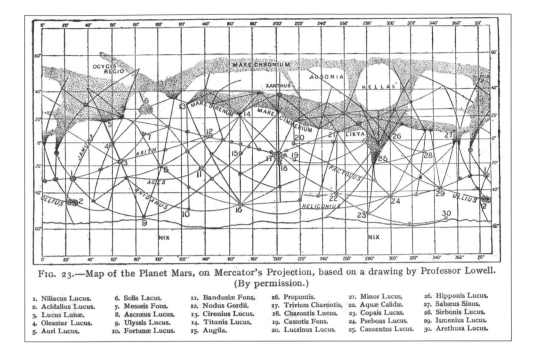

FIG. 23.—Map of the Planet Mars, on Mercator's Projection, based on a drawing by Professor Lowell. (By permission.)

1. Niliacus Lucus.	6. Solis Lacus.	11. Bandusiæ Fons.	16. Propontis.	21. Minor Lucus,	26. Hipponis Lucus.
2. Acidalius Lucus.	7. Messeis Fons.	12. Nodus Gordii.	17. Trivium Charontis.	22. Aquæ Calidæ.	27. Sabæus Sinus.
3. Lucus Lunæ.	8. Ascræus Lucus.	13. Cirenius Lucus.	18. Charontis Lucus.	23. Copais Lucus.	28. Sirbonis Lucus.
4. Oleaster Lucus.	9. Ulyssis Lucus.	14. Titanis Lucus.	19. Cassotis Fons.	24. Pseboas Lucus.	29. Ismenius Lucus.
5. Auri Lucus.	10. Fortunæ Lucus.	15. Augila.	20. Lucrinus Lucus.	25. Casuentus Lucus.	30. Arethusa Lucus.

31   Mars of the canals: Lowell's standard map as used by Flammarion in his Astronomy for Amateurs, *1903*.

today'. The confusion grew: two prestigious observatories in the USA found in one case no canals, in another a few of them but no geminations, and no changes to Libya. From the Harvard University southern hemisphere observatory located at Arequipa, Peru, William Pickering reported clouds in the Martian skies, and large lakes spread out beneath them. The Lick Observatory in California with a 36-inch refractor saw none of these features; perplexity spread.

Whilst observers exchanged formal reports and papers, the popular writers Flammarion and Proctor were quickly at work. Proctor thought the canals might be rivers; he was amongst the first to note that a Martian canal would have to be 'fifteen or twenty miles broad' to be seen from Earth, at least with the telescopes available at the time. Flammarion, greatly heartened by the Brera discoveries published his definitive *La Planète Mars* in 1892: 'the canals may be due... to the rectification of old rivers by the inhabitants for the purpose of the general distribution of water...' (**29**). Other commentators supposed that the 'canals' might be optical illusions, an argument first advanced by the English artist Nathaniel Green, teacher of painting to Queen Victoria and an amateur astronomer.

The canals debate might have levelled off at this point had it not been for the incursion of its most prominent controversialist, the convinced pluralist, Percival Lowell. Lowell, an eminent Bostonian, entered the astronomical fray after a career in business and diplomacy, mainly in the Orient. He was a formidable mathematician, a master of the detached, ironic prose fashionable at the time, and immensely wealthy. He financed his own observatory

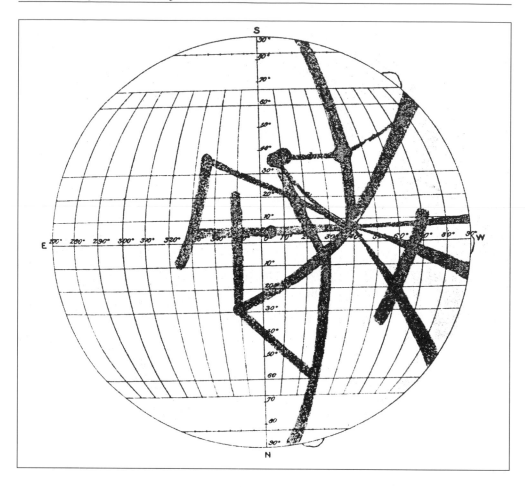

32   Canals on Venus? Lowell's drawing of Venus, October 19, 1896. Astronomers feared 'canalism' was going to infect other parts of the solar system although Lowell actually doubted that the markings were of artificial origin; he made similar yet more complex drawings of Mercury, 1897.

at Flagstaff, Arizona and commenced work in 1894 (**30**). He chose the site for his work after carefully scouting around for a location that would yield clear, steady skies; in astro-parlance, good 'seeing'. In this matter, especially regarding the necessary steadiness of the atmosphere, which is now a commonplace consideration in the siting of observatories, he was a pioneer. He may not, however, have brought an entirely objective mind to the task of observing Mars. Even before he started operations he had announced that the canals were probably 'the work of some sort of intelligent beings'.

Lowell was indefatigable. In addition to writing numerous papers and lecturing on Mars and its canals, he wrote three readable best-sellers: *Mars* (1895); *Mars and its Canals* (1906); and *Mars as an Abode of Life* (1908). These, and his last major work *The Evolution of Worlds* (1909) illuminate the philosophical background against which his Martian work should be perceived — firm evolutionism. The newly-arrived popular press was very willing to

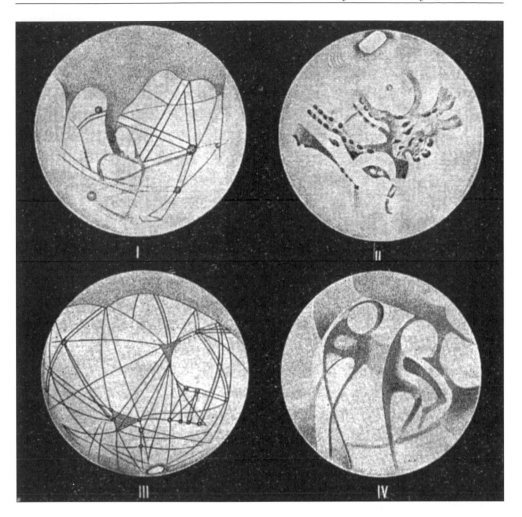

33   In the eye of the beholder: four representations of Mars, l to r: Schiaparelli, 1877; Antoniadi, 1909; Lowell — also 1909; and from photographs taken at Pulkovo Observatory, Russia, 1909.

report his findings and views; canal mania grew apace. By 1910 Lowell had reported over 400 canals with an average length of 1500 miles each (**31**). He wrote plausibly about a Martian atmosphere and the means by which the canals distributed water from the polar caps in order to irrigate the planet before evaporation returned water to the poles. Lowell reported that there was a seasonal darkening around the canals, possibly freshly irrigated land sprouting vegetable growth.

This hydraulic cycle appealed to popular evolutionism which, influenced by Proctor and Flammarion amongst others, liked to perceive Mars as an old, dying world trying to avert its fate by rational and cyclopean engineering projects; this was, after all, an era of great canals: Panama, Dortmund-Ems, Manchester, Corinth.

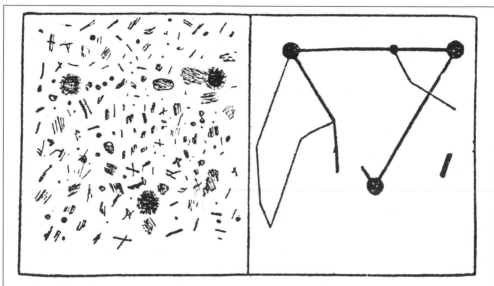

IS PROFESSOR LOWELL DECEIVED ?

This diagram bears out the contention of some critics that what Professor Lowell thinks canals are illusions. Hold this 20 feet away, and the marks on the left look like the lines on the right.

34 An optical illusion? The chief weapon of Lowell's critics including Green and Maunder, here illustrated in Harmsworth's Popular Science *(1912).*

For all his enthusiasm and public popularity, Lowell remained in a small minority of hard-line canalists. Like Wright or Kant, he was not an astronomer's astronomer. Andrew Douglass, one of his assistants at Flagstaff (and who had assisted Pickering at Arequipa) had doubts about his chief's objectivity, about his tendency to find facts to illustrate a theory rather than vice-versa; he was sacked for his pains. Lowell worked on; he noted web-like patterns on Mercury and Venus and, although he interpreted them cautiously, the astronomical world feared that canalism was going to infect other parts of the solar system (**32**). A series of personality clashes at Flagstaff (not directly involving Lowell, however) and general stress resulting from overwork brought about nervous prostration; Lowell was *hors de combat* 1897–1901, but then returned to the front line fighting hard for Mars of the canals. His energy and ingenuity were legendary; he married later in life and on his honeymoon in London in 1908 he went aloft with his bride in a balloon ascending to 5,500 feet, in order to observe whether the roads and paths traversing Hyde Park gave a canal-like reticular pattern when seen from afar.

Ironically, Lowell's main critic and the ultimate undoer of his elaborate theories, the Greek astronomer Eugene Antoniadi was himself a pluralist who thought at one time that he had seen some Martian canals. But this accomplished observer became convinced that the canals were indeed optical illusions and the majority of the astronomical community agreed with him (**33**).

Phantasiebild des Querschnitts durch eine mit einem „Kanalsystem" versehene Talsenke (nach
Schiaparelli, 1895): An den Abhängen terrassenartig übereinander angeordnete Kanäle. An beiden
Enden des Talzuges Staudämme mit Schleusen. Man füllt m und m', bis sie überfließen und
die Abhänge bis n und n' bewässern; in gleicher Weise läßt man nach Bedarf und Möglichkeit
die Abhänge bis p und p' und die Talsohle in der Bewässerung folgen.

35   *Public sector canals; a diagram from Robert Henseling's* Mars *(Stuttgart, 1925) depicting
Schiaparelli's theory of differentiated Martian canal levels. In Spring, 'The Minister of
Agriculture orders the opening of the most elevated sluices ...' on a collectivist Mars.*

The doyen of the staff at Greenwich Observatory, Edward Maunder, carried out an
experiment at the nearby Royal Hospital School in 1903 whereby boys were asked to
record their impressions of prepared diagrams presented to them. Their responses
suggested that canal-like patterns could be perceived and generated by the human mind
even where they did not exist. This early gestalt experiment failed to convince Lowell who
pooh-poohed it as the 'small boy theory', but it cut ice with the scientists who respected
systematic enquiry (**34**).

In December 1909 Antoniadi spoke for many other astronomers when he effectively
wrote finis to the Martian canals: 'Under good seeing there is no trace whatever of a
geometric framework...'.[4] The Spanish astronomer J. Comas Solá concluded: 'The
marvellous legend of the Martian canals has disappeared'. On even safer ground, the
American astronomer E.B.Frost on being asked by a newspaper for a 300-word article on
the 'habitability of Mars' responded: 'three hundred words unnecessary — three enough
— no one knows'.

Psycho-history has analyzed Lowell in recent years, although his motives still remain a
mystery. Why did he take to astronomy on such a scale? What did he see or think he saw
on the surface of Mars? To some people he is the Brahmin disillusioned with war, coarse
materialism, nationalism and excitable democracy. He liked, perhaps, to imagine sensible
Martians organising their planet rationally, an argument which clearly had wide popular
appeal. C.K.Hofling[5] thought Lowell was influenced by 'voyeuristic impulses...
unresolved oedipal conflicts...'; who shall say?

As is often the case it was a prophet's acolytes who pushed his propositions *ad absurdum*.
C.E.Housden, a British hydraulic engineer in his *The Riddle of the Planet Mars* (1915, one
year before Lowell's death) described an elaborate reticule of Martian canals: single,
double, two-way, oasis junctions, varieties of pumping stations, and so on. This, and other
ingenious explanations and amplifications failed to resuscitate the fast-fading canal theory
(**35**). Although it had adherents as late as the 1950s, space probes and Mars landings
eventually killed off the vestiges of Lowell's grand theory.[6]

Lowell had taken pluralism, and especially popular pluralism, to undreamed-of heights.

Drawn by M. Wicks

36   *The city of Sirapion, Mars from Mark Wicks' Lowellite novel for young readers* To Mars via The Moon *(1911) – the travellers to Mars depart from and return to 'Norbury in the County Borough of Croydon.' In the story the Martian men wore Balkan apparel, the women 'long hose and tunics'.*

The failure of the canal hypothesis left pluralism embarrassingly discredited. It had hitched its wagon to the prestige of science and science had revealed one of its major claims to be badly wanting. Carl Sagan, an eminent twentieth-century pluralist, suggested that the calamitous failure of the canal episode held back planetary astronomy for years. It is true that astronomy turned to deep space soon after the canals fiasco, but developments in theoretical physics and instrumentation contributed strongly to this shift of emphasis.

Lowell's flair for publicity nevertheless helped to launch one of the most popular cultural icons of modern times: fictional Mars, rich with promise for creative intellects (**36**). Prior to the canal episode, novelists had described an inhabited Mars, as well as other celestial bodies. The Moon, for example, had been the setting for fiction and satire from

37   A 'disorder of rocks...bristling scrub...I floated through the air and fell like a feather' — Claude
      A Shepperson's stylish rendering of the initial steps of H.G.Wells' The First Men on the
      Moon, 1901.

the time of Lucian; over 150 works of 'lunar fiction' of one kind or another had appeared
by 1790[7] (**37, 38, 39**) and it continued to be a popular location for another century.

A red and pleasant land

Fictional Mars had hitherto enjoyed a more modest bibliography. One author whose work
spanned the pre- and post-canal eras was the polymath Rev Wladislaw Lach-Szyrma, vicar
of Newlyn, Cornwall.[8] Lach-Szyrma's Mars, described in his *Aleriel, or a Visitor from Other
Worlds* (1883) was inhabited by a well-evolved leonine race of great technical ingenuity and
moral worth, evidently the beneficiary of some kind of redemption from an earlier, fallen
state. Lach-Szyrma, a keen astronomer and an advocate of the plurality of worlds heard
about the canals at an early stage and worked them into his stories; huge floating cities (**40**)

38 *A dead civilisation on the Moon; Angus Macdonnall's fine illustration for John Ames Mitchell's* Drowsy, *1917, a late date for this recurring pluralist theme.*

moved about on the surface of his Martian waters drawing sustenance from aquatic life.

One-third of late Victorian and Edwardian space fiction was located on the 'Red Planet'.[9] Some of it set enduring standards. Probably the best-known was H.G.Wells' *The War of the Worlds* (1897). One of the most famous openings in English literature tells the reader that 'in the last years of the nineteenth century... human affairs were being watched keenly and closely by intelligences greater than man's yet as mortal as his own'. Wells had a range of sub-texts: ruthless, indifferent evolutionism; imperialism able to command superior technology; intellectual and moral development out of kilter; all abiding themes in any modern society (**41**). Lowell's Mars supplied Wells with a perfect source, albeit not the kind Lowell had in mind.

Imaginary Mars had a promising affinity with terrestrial fashions and ideologies. Edgar Rice Burroughs, later to invent Tarzan, used the planet as a setting for traditional folk-fantasy, Mars inhabited not only by monsters but also by egg-laying princesses of great beauty. Kurd Lasswitz described enlightened Martians raising moral sights on Earth in Germany's pioneering space fiction novel, *Auf zwei Planeten* ('On Two Planets', 1897). In a more

39   *Gold mining in the
       mountains of the Moon:
       an imaginary scene by
       Fred T. Jane (of* Jane's
       Fighting Ships*) —
       Pall Mall Magazine,
       1894–95.*

austere vein, the Bolshevik Aleksandr Bogdanov wrote *The Red Planet* (1909) in which 'the democratic state was forced to involve itself with the [canal] project in order to absorb the growing surplus of the proletariat and aid the remnants of the dying peasantry'.

Early film makers seized on the idea of an inhabited Mars for some of their productions. By 1914 there were half a dozen such films including a pioneer New Zealand production (*A Message from Mars*, 1909) and a British film of the same title (1913) in which a Martian comes to Earth to reform a fallen soul, an implied reversal of Chalmers' vision of a century before. Perhaps most poignant of all these early Mars films was the pioneer Danish space-opera *Himmelskibet* (Heaven Ship) in which a daughter of the Martian high priest, white-robed, vegetarian and pacifist, marries a space-traveller from Earth and returns with him to plea for peace and reason in a world fallen prey to war. The film (**42**) was made in 1917, the year of the U-boat blockade, Third Ypres and the Russian Revolution. Lowell's Mars and the controversies surrounding it were by now lost in the preoccupations of the four years' holocaust.

40 *Canal city: Paul Hardy's faithful representation of Lach-Szyrma's floating city sailing along a Martian canal (1890) — the Statue of Liberty-esque figure with raised arm, left of centre, is a memorial to a Martian 'female freedom fighter'.*

41 *Martian touch-down, Surrey: Victor Reinganum's illustration for the* Radio Times *(1950) — the artwork and radio programme were closer to Wells'* The War of the Worlds *than most attempts. (Courtesy:* Radio Times*).*

42   *Early space-opera: the 1917 Danish film* Himmelskibet *(Heaven Ship) — the daughter of the pacifist, vegetarian Martian high priest returns to enlighten war-torn Earth. (Courtesy: A/S Nordisk Films Kompagni, Copenhagen)*

References

1   J. E. Filachou, *De la Pluralité des Mondes*, 1861.

2   Mars' 'opposition' to Earth when it lies on the opposite side of the Earth to the Sun can vary from 35 to 63 million miles.

3   Michael J. Crowe, *The Extraterrestrial Life Debate 1750-1900*, 1986.

4   British Astronomical Association Journal 20, Dec 1909.

5   C. K. Hofling, 'Percival Lowell and the Canals of Mars', *British Journal of Medical Psychology*, 1964.

6   There remains the unfalsifiable explanation that the canals were indeed there on the particular occasions that Lowell observed them.

7   Marjorie Hope Nicolson, *op cit.*

8   Lach-Szyrma's father was an eminent Polish exile; his mother the daughter of one of Nelson's captains. See: R. A. S. Hennessey: 'Nineteenth Century Mars, a Cornish Connection', *Foundation* 68, Autumn 1996.

9   G. W. Locke: *Voyages in Space: A Bibliography of Interplanetary Fiction, 1801-1914*, 1975.

10 The quiet years, 1920-1950

Pluralism lying low

By most historical criteria the period from the end of the Great War to the 1950s would be perceived as anything but tranquil; the years of the Great Depression, crash industrialisation and terror in the Soviet Union, the horrors of the Holocaust and a Second World War. Radio, cinema, and motor transport spread disorienting social change rapidly. Pluralism, however, remained becalmed. Its spirit was kept alive by burgeoning science fiction, but as a mainstream intellectual pursuit it seemed *démodé*. The spirit of the times was unsympathetic to it; cultural experiments, new fashions and new concerns tended not to relate to the grand metaphysical traditions which had sustained it for so long: few ideas lack more lustre than those whose day has recently passed.

The late Victorians who had linked pluralism so hopefully to science now found that the sciences were weakening rather than supporting it. The first major attempt at scientific verification, the saga of the Martian canals, had ended in tears. By the 1920s the canals were generally dismissed as illusory, although they still had a few supporters. William Pickering and Earl C. Slipher, both Flagstaff alumni were loyal to Lowell's vision; Slipher asserted (1921, and again in 1931 and 1940): 'Photographs of Mars... furnish objective evidence... of the existence of dark, linear markings — the "canals"....' Although popular fiction continued to cherish the canals, science had abandoned them.

Instead, astronomers came to favour a theory about the origin of the solar system which was distinctly disobliging to pluralism. This theory, first put forward by T.C.Chamberlin and F.R.Moulton in the USA (1901) and later elaborated by Sir James Jeans in Britain (1916), was that the solar system was the result of an outside chance: the Sun and another star had once passed close to each other, had even 'grazed', and the resulting debris had eventually coalesced into planets. If this were so, solar-type systems were quite possibly a great rarity in the Universe, as were the intelligent life forms that had appeared in one of them. Chamberlin put it thus many years later: 'Our planetary family had an aristocratic birth'. Jeans put it more starkly: 'All this suggests that only an infinitesimally small corner of the universe can be in the least suited to form an abode of life.'[1]

At first it appeared that biology was also administering a cold shower to the pluralist case. Alfred Russel Wallace, co-discoverer of biological evolution with Darwin, had died in 1913, but his writings remained popular for another decade. His main anti-pluralist tract *Man's Place in the Universe: A Study of the Results of Scientific Research in Relation to the Unity or Plurality of Worlds* (1903, with eight editions up to 1914) drew heavily on evolutionary theory, but also from the author's metaphysical hypothesis that 'some higher intelligence may have directed the process by which the human race was developed'. Like Whewell, Wallace had an anthropocentric distaste for the implications of pluralism. He also drew extensively on astronomy, employing the assumption, not uncommon in his day but very shortly to be shown as false, that the Sun was situated at the centre of the Milky Way system, a galaxy which he assumed, also wrongly, constituted the entire Universe. His views carried weight and, more important for the future, they raised the biologist's question: what is the actual nature of life? Wallace suggested that the evolution of life was the result of a highly complex process requiring vast tracts of time: 'it seems in the highest degree improbable that [the causes of life] can all be found combined either in the solar system or even in the stellar universe'.

The Wallace theory had a mixed reception. Camille Flammarion was predictably unconvinced by its 'geocentric and anthropocentric' pleading; other leading astronomers also expressed their doubts. Within twenty years the 'new astronomy' which was informed by Einstein's theory of relativity and the findings of the Mount Wilson 100-inch telescope, could demonstrate Wallace's errors. By 1925 it was clear that the Sun and its system were well off-centre in a galaxy which was but one of many in a far vaster Universe than anyone had previously imagined. This was better news for pluralism, but it took a long time for its implications to be realised.

Signals to and from space

In spite of the uncongenial climate for pluralism in the inter-war period, and its generally unexciting record, the pluralist chronicle was not altogether empty. Developments were in train which were eventually to carry the new, science-based pluralist ideas farther than their nineteenth-century advocates could have imagined. These initiatives took a long time to reach levels where they could offer practical assistance to pluralists: radio astronomy; rocketry; and particular aspects of biology and biochemistry. Nevertheless they were well under way by the mid-1930s.

The idea of signalling to the Moon and to intelligent extraterrestrials was already about a century old. As technology developed, and pluralist aspirations became bolder, so space-signalling suggestions had grown in number and scope (**43, 44**). Camille Flammarion had enthusiastically backed his fellow countryman Charles Cros when he suggested to the French Academy of Sciences (1868) that Mars or Venus could be signalled by means of large parabolic mirrors reflecting electric arc lights. One of Flammarion's considerable female readership, Mme Guzman, was moved to bequeath 100,000 francs (then worth about £4200) for a prize in memory of her son 'for the person who will find the means within the next ten years of communicating with a star (planet or otherwise) and of receiving a response'. Flammarion was sanguine about the outcome; if anything, signalling planets was 'perhaps less bold' than the telephone, the phonograph or the 'kinetograph' (a precursor of the cinema).

43 Commercial Possibilities: projecting advertisements on to the Moon; the great searchlight of the 'Ashwanipi Falls generator' approaching meltdown, George Allan England's story in Pearson's Magazine, *1907; artist, Warwick Goble.*

There were other signalling plans. In the 1890s the Rev Lach-Szyrma thought that lighting up the Rigi, or the snow-covered Malvern Hills, with electric beams in geometric form might be worth a try, or even illuminating a huge, floating cross on Lake Michigan, powered from the Chicago Exhibition (**45**). Francis Galton, a meteorologist, worked out a possible 'effective interstellar language' by describing simple mathematical exchanges resting on a base 8 notation. The pluralist astronomer William Pickering took the matter up during the 1909 opposition of Mars, towards the end of the canals affair; for $10 million there could be an array of 5000 10ft mirrors flashing signals to Mars. Other ideas of those optimistic years included the regular folding and unfolding of black cloth strips in the desert, or legions of cut-price hand-held mirrors operated manually from the plains of Texas.

Two signalling plans were pregnant for the future. In 1899 the banker J.P.Morgan financed the wayward genius of early electricity, Nikola Tesla, to build a radio transmitter at Colorado Springs. Employing a 75ft coil Tesla not only sent the first radio message to Mars but also, so he claimed, received a similar one in return: probably low-frequency

44   *Signalling Mars: (upper) the
      favoured late Victorian and
      Edwardian approach, by
      searchlight;
      (lower) cherubic Martians gaze
      at Earthly life; (*London
      Magazine, *1907).*

THE FIRST SIGNAL FROM LAKE
MICHIGAN.

45   Signalling extraterrestrials: Lach-Szyrma's proposal for signalling Mars by means of a cross of arc-lights floating on Lake Michigan; Cassell's Family Magazine, *1893.*

waves caused by lightning discharges. Also, during the 1909 excitements, Professor David Todd of Amherst College planned to set off in a balloon with a radio receiver listening for 'Herzian waves', Martian signals, to Earth. But his scheme hit a now-familiar snag: there were already some 2000 radio transmitting stations on planet Earth; their signals would, it was feared, interfere with Todd's experiment.

During and after the Great War, Todd was still keenly at work hoping to repeat his listening-in attempt and also to transmit to Mars with the assistance of the balloon experts of the US Army Signal Corps at Omaha, Nebraska as well as 'experts from the Rockefeller Institute'. This scheme also bore little fruit, only some ground-based attempts at communication. The opposition of Mars in 1924 gave Todd a further chance for experimentation. Amazingly, he persuaded the US Army and US Navy to maintain periods of virtual radio silence for two days and to alert him about any messages from Mars that they might pick up. At the same time British radio specialists at Dulwich thought they had picked up a coded space-signal (**46**).

The pioneer of wireless telegraphy Guglielmo Marconi also took a close, if fitful, interest in signals from Mars. He announced in 1919 that he had occasionally received unexplained signals and that he hoped for 'communication with intelligences on other stars'. The *New York Times*, however, warned that humanity should 'leave the stars alone', particularly since it was unprepared for messages from 'superior intelligences' — an early

THE " CANALS " OF MARS.

Factors Needed for their Observation.

REPLY TO THE DOUBTERS.

(By Our Astronomical Correspondent.)

The statement often made that the *canali* of Mars may be a purely subjective phenomenon and have no objective existence is quite erroneous. It is true we are doubtful about their real appearance—their width and their continuity—but we have no doubt whatever that they have a real existence. Those who make this statement urge, in support of it, that the majority of astronomers have never seen the *canali*, and photographs taken with some of the largest telescopes show no sign of them. That is perfectly true, but it has little or no bearing on the problem of the *canali*.

The majority of astronomers have never seen the *canali* for the simple reason that the majority of astronomers do not observe Mars. The study of the planets, and of Mars in particular, forms a small and comparatively unimportant branch of astronomy. The majority of astronomers—practically all of those who work at public observatories—are not concerned with it: their entire time is occupied in the photographic and spectroscopic examination of the sun and stars, in the measurement of the positions of the heavenly bodies, and in mathematical investigations. Planetary astronomy is left, almost entirely, to "amateur" astronomers and their assistants.

IS IT A MESSAGE FROM MARS?

Mysterious Marks on a Scientist's Film.

(From Our Own Correspondent.)

WASHINGTON, August 26.

Dr. David Todd, Professor Emeritus of Astronomy at Amherst College, whose studies of Mars have extended over many years, has just shown me a film which his friend, Mr. C. F. Jenkins, kept running from last Friday noon until Saturday at five o'clock in the afternoon at his laboratory in this city. It was Professor Todd's theory that if Mars is inhabited and the Martians attempted to signal the Earth it would be only possible to receive their signals by the conversion of light values into electric values.

He therefore persuaded Mr. Jenkins to set in motion an intricate piece of mechanism of his own invention, which may be briefly and untechnically described as a light-proof box, containing a slow-moving film, practically the same as a film encased in an amateur's ordinary camera, but with this difference, that whereas in an ordinary camera light passes through the lens to the film behind it and produces an image, in Mr. Jenkins's camera the lenses are within the box and radio waves caught from ether pass through the lenses inside the box and thence to the film. Instead of the film being turned by hand from one reel to the other, as the amateur does after making the exposure, in Mr. Jenkins's machine the film runs by clockwork and was kept in constant motion for nearly 30 hours. Placed in a developing bath and developed in the ordinary way the film shows distinct marks to the naked eye. Under the microscope they are revealed as blocks of dots, some of the dots larger and blacker than others. These blocks occur at regular intervals on the entire strip of film, about 25ft. long.

46   *SETI, 1924 style: extract from* Morning Post *reports in August 1924 suggesting the possibility of Martian light-signals, and an interesting twist in late 'Canal Era' debating tactics.*

rehearsal of a division of opinion that was to reappear later in the century. Marconi was still claiming that he had possibly tuned in to signals from Mars whilst on his yacht *Electra* crossing the Atlantic in 1922, but thereafter his interest cooled. To one of radio's chief theoreticians, Charles Steinmetz, all these experiments were so many pinpricks; if humanity seriously wished to signal Mars it had to face the construction of massive radio beacons and the generation of equally staggering quantities of electric power. The combination of an uncertain outcome and the extreme unlikeliness of sufficient funds forthcoming for sustained experimentation discouraged further development; signalling to, or from, other worlds faded in the generally subdued pluralist scene of the times.

But not entirely. In 1931, the Rev E.W.Barnes, Bishop of Birmingham and a scientist, took what was for that era a bullish line on pluralism. He 'had no doubt that there are many other inhabited worlds', the 'star-grazing theory' of planetary evolution notwithstanding. He therefore advocated a more ambitious extraterrestrial communications programme, tuning in to distant stars rather than planets in Earth's backyard. Shortly after Barnes made his suggestion the first intimation of a means of executing it was being developed by Karl Jansky, a radio engineer working in New Jersey on the problem of static interference. He discovered the narrow band of wavelengths that convey radio messages to the Earth from outer space; the *New York Times* put it neatly: 'New radio waves traced to centre of Milky Way.' The full potential of Jansky's discovery was only exploited a generation later; nevertheless more seeds of pluralism's eventual renaissance had been sown.

Rockets into space

There has long been a close connection between rocketry and pluralism. Many of the pioneers of modern rocket science had pluralist leanings; they belonged to a tradition that coupled thoughts about life on other worlds with 'the possibility of a passage thither' as John Wilkins had put it in 1638. Early fiction concerned with pluralism abounded in space-travelling devices.

Bishop Godwin's 'Man in the Moone' was borne through space clutching a species of sail-guided trapeze held aloft by 'gansas', unusual geese or swans. Cyrano de Bergerac, having tried bottles of dew (which might rise in the Sun's heat) settled for a rocket-powered chariot (1657); Kindermann's travellers set off for Mars in a spaceship lifted by six evacuated copper globes. A wide range of projectiles flung from the Earth by guns, springs or centrifugal force adorned Victorian space literature, none of them practical.

In Russia, K.E.Tsiolkovsky (1857–1936) a provincial teacher in Kaluga was attending to the problem more scientifically (**47**). Now perceived as 'the father of space flight' Tsiolkovsky worked at that time in obscurity. Pluralism was one of his inspirations; he was amongst the advocates of signalling Mars by means of mirrors, reported by the *Kaluzkskii Vestnik* ('Kaluga Herald') in 1896. In order to popularise his highly theoretical work on rocket design he wrote descriptive science fiction including *Dreams of Earth and Sky* (1895) and *Outside the Earth* (1916). In these and other works he speculated about the forms extraterrestrials might take, for example contrasting the relative advantages and disadvantages of dwarfs and giants in given gravity conditions.

47 *Konstantin
    Eduardovitch
    Tsiolkovsky; Russian
    pluralist and pioneer
    of computations for
    spaceflight.*

Tsiolkovsky's star rose after the Bolshevik Revolution in 1917; the new regime, not unsympathetic to pluralism, was anxious to establish its modernist credentials and spaceflight was considered to be a promising candidate for this. Lenin had pluralist thoughts: 'if we succeed in establishing interplanetary communications, all our philosophical, moral and social views will have to be revised' — hardly the kind of sentiment one might expect from his contemporaries in the UK or USA, Stanley Baldwin or Calvin Coolidge. In 1929 the USSR published a nine-volume encyclopedia of space engineering: *Interplanetary Communications* which devoted an entire volume to the works of Tsiolkovsky; the Soviet state also financed a Gas Dynamics Laboratory and, in the era of the rising acronym, GIRD (Group for the Study of Reaction Propulsion), both organisations devoted to rocketry and its implications for spaceflight.

No other state supported the theory and practice of spaceflight to a similar extent. Private enthusiasm nevertheless accomplished a great deal. In Weimar Germany a band of enthusiasts formed the *Verein für Raumschiffahrt* (Society for Space Travel) two of whose early members had pluralist sympathies: the leading theoretician Hermann Oberth, later to express the view that unidentified flying objects were 'spaceships from another solar system', and the practical Wernher von Braun, a rocket constructor with a taste for speculative science fiction, more interested in space travel than the war missiles he was later obliged to design by the Nazi government.

In the USA the American Interplanetary Society (later the American Rocket Society), contained some committed pluralists; David Lasser, president of the AIS wrote a popular book advocating space travel, *The Conquest of Space* (1931) in which he devoted a chapter to pluralist ideas, to the 'peaceful conquest of our sister worlds'. He concluded after a well-informed survey, that Mars and Venus 'may support life of their own of the order of terrestrial life' and, quoting Flammarion, warned the reader not to be unduly influenced by the imperatives of terrestrial life support systems: 'There is a possibility that sulphur and silicon and other elements can well serve for the bases of other life forms.'

The British Interplanetary Society (BIS, founded in 1933) which also advocated space exploration, ran on a shoestring whilst some of its members kept the British pluralist vision alive. Two of them, the science fiction writers William Temple and Arthur C. Clarke managed its operations from a small flat in Gray's Inn Road, London, 'trying to convince a sceptical world that men would one day travel to the Moon'.[2]

Comparisons between the outcomes of these private initiatives are instructive. The German rocketeers were conscripted by the Nazi state for war work. The American effort, which rested as much on a lone pioneer, Robert Goddard, as on limited private enthusiasm, became ultimately the most productive and prestigious on Earth once it was taken under the wing of the US government after 1945 — borrowing heavily from German expertise. But when one of the founder-members of the BIS took up the matter of rocket development with the British government he received dusty answers; experimentation with liquid-fuel rockets, where the future was to lie, might 'contravene the provisions of the Explosives Act of 1875', an augury of the rather unenthusiastic vacillation which was to characterise British space policy in later years.

Science fiction

Although the intelligentsia was disenchanted with pluralism between the world wars, the spirit and achievements of earlier pluralists were honoured and kept alive by the writers and readers of science fiction. Its range of quality was wide, from some of the loftiest pluralist speculation ever expressed in fiction in the works of Olaf Stapledon, to the fast-growing field of pulp magazines like *Amazing Stories* (1926) and *Astounding Science Fiction* (1930). Magazines such as these popularised the notion of grotesque, often threatening, extraterrestrial creatures which, because of the lurid illustrations which often accompanied the stories, introduced the clichés of the 'BEM' (Bug-Eyed Monster) and 'LGM' (Little Green Men) into everyday metaphor (**48**).

The magisterial *Encyclopedia of Science Fiction*[3] whilst noting that science fiction only accounted for some 2-3% of the total pulp market, observed: 'Early US pulp-magazine sf in the vein of Edgar Rice Burroughs populated other worlds with quasi-human inhabitants — almost invariably including beautiful women for the heroes to fall in love with — but frequently placed such races under threat from predatory monsters', a neat

48   *Mercurians: Virgil Finlay's illustration for John Hawkins' 'Ark of Fire', in* Famous Fantastic Mysteries *(1943)*.

summary of a standard story-line (**49**). An important feature of pulpdom was that its products, especially offshoots like the space-fiction comics which featured Buck Rogers or Flash Gordon and their cinematic derivatives, were enjoyed by young people. Although it would be difficult to prove a hard cause-and-effect chain, this lively fictional propaganda may well have sown seeds for the postwar renaissance of pluralism, at the very least by raising consciousness, and implanting questions (**50**).

The Liverpool philosopher Olaf Stapledon, on the other hand, produced literary masterpieces placed squarely in the 'great tradition' of philosophical pluralism, sustaining it in unpropitious times. His major pluralist work, *Star Maker* (1937) is a dream-like navigation of a cosmos replete with life. Brian Aldiss described its range: 'human echinoderms, and intelligent ships... symbiotic races, multiple minds, composite beings, mobile plant men, and other teeming variants of the life force. Utopias, interstellar-ship travel, war between planets, galactic empires... telepathic sub-galaxies going down in madness....'[4] C.S.Lewis was another thinker sympathetic to pluralism. He also used an

49   *Pulp Art: illustration by science fiction artist (Frank) Paul for 'The Return from Jupiter' by Gawain Edwards (the rocket engineer Edward Pendray);* Wonder Stories, *March 1931; the 'angels with knives' are another variant of anthropomorphic alien.*

other-worldly background to explore questions of morality and spirituality, for example in the interaction between humans and weird Martians such as the 'hrossa', giraffe-like beings who spoke fluently and who were, therefore, set firmly in the long tradition of extraterrestrials whose characteristics raised difficult questions about the ways of humans back on Earth (*Out of the Silent Planet*, 1938).

The perceived gap between science fiction and other literature (more pronounced in Britain than in mainland Europe and the USA), the apparent mind-set of prestigious mainstream science against pluralism, and other preoccupations of the time, all combined to restrict pluralist sympathies, even of a passing, intelligent interest in the subject, to a minority of people.

50   *The Trans-Uranian Railway: having failed to persuade the authorities of planet Earth to build his monorail, Chalmers Kearney festooned Uranus with it in his novel* Eróne *(1943), 'full of sound sense' according to George Bernard Shaw.*

Biology, physics and metaphysics

In spite of occasional pronouncements and Wallace's strong intervention on the anti-pluralist side, the pluralist potential of biology had been but lightly tapped. The work of two biochemists during the 1920s, A.I.Oparin in the USSR and J.B.S.Haldane in Britain concerning the origin of life had eventual pluralist implications. They arrived independently at the hypothesis that original, primitive life may have started in some kind of primordial 'soup' when the Earth was young and hot. At first Oparin was suitably prudent on the question of whether or not such soups might have been concocted on other worlds, particularly when the prevailing paradigm in the cosmogony of the solar system was Jeans' 'star-grazing theory'. In his *Origin of Life* (1936) Oparin ventured to say: 'We still have too little information to deny completely the possibility of existence of organisms on some other planets, whirling about stars similar to our Sun....' Later in life he became more intrepid, perhaps as a Soviet citizen influenced by Marxism's accommodating attitude to pluralism. By 1953 the title of his book had expanded: *The Origin of Life on Earth*, and with it he suggested that life might even now be starting on other worlds; three years later he collaborated on a further book — *Life in the Universe*.

Haldane, who read Stapledon's fiction, was easier-going than Oparin towards pluralism. In an essay 'The Last Judgement' (within *Possible Worlds*, 1927) he outlined, in fictional form, some grim possibilities regarding the Venus of the future where only eleven out of 1,734 spacecraft had landed successfully. Their crews discovered organisms there but, finding that they were unsuitable as food and possibly dangerous to humans, Earth scientists undertook a lengthy programme to isolate bacteria 'synthesised ... to attack l-glucose' which duly wiped out the native, Venusian organisms. The possibilities of a new Venus could now be unfolded: 'the evolution of the individual has been brought under complete social control...'. Human fears about malevolent extraterrestrials are hardly surprising after this expression of cosmic genocide.

51 Nobel pluralist: the Swedish scientist Svante Arrhenius (1859-1927) — advocate of the panspermist theory, and of life on a lush, sweltering Venus. (Courtesy: Royal Swedish Academy of Sciences)

In its heyday and for some time thereafter the Oparin-Haldane theory helped to keep open vital pluralist questions, although experiments in 1953 suggested that amino acids, the basic building blocks of life, could be synthesised in artificially-created 'primordial soup' conditions, many other necessary ingredients for living cells were not so created.

Astronomy offered rather more hope for pluralism, but of an exceedingly cautious kind in the wake of the Martian canals controversy. Although the star-grazing theory dampened pluralist hopes, this was to a small extent offset by a series of observations and discussions about Mars from 1924 onwards. The gist of these was to conclude that Mars might indeed bear life, but most possibly low-grade vegetation, and certainly not sentient canal-builders. The *New York Times Magazine* (1928) found that a majority of 'Eminent Astronomers' went along with the idea of Martian vegetation and even, in a few instances, of simple fauna. This hypothesis lasted until the 1950s but it remained down-beat: the leading experts on the question of Martian life, Gerard Kuiper in the USA and G.A.Tikhov in the USSR conceded that there were probably lichen-type growths on Mars.

There were also occasional exchanges about life on Venus. The Swedish Nobel Laureate, Svante Arrhenius (**51**), suggested the planet was a sweltering hothouse: 'the average temperature is calculated to be about 47°C... everything on Venus is dripping wet'.[4] Arrhenius' vision of a planet of smothering cloud layers, luxuriant vegetation and massive swamps was offset by that of F.E.Ross: of an arid Venus wracked by immense dust storms, swirling under eternal cloud banks. In 1920 the American astronomer Charles Abbott thought, however, that Venus was 'a twin planet to the Earth... lacking in no essential to habitability', it was a good candidate to replace discredited Mars as a possible seat of life. In 1926 he dared to wonder about the possibilities of intelligent life on Venus, of 'fluent communication by wireless with a race... having their own systems of government, social usages, religions...', although such astronomical research as there was on the subject gave no grounds for these musings.

Popular astronomy books still had short passages on pluralism, most often confined to comments on Mars and the canals theory, and usually treating its outcome as problematic. Jeans thought 'on the whole the case for life existing on Mars, or any other planet of the solar system, can hardly be called a strong one'; George Forbes thought it unlikely that 'on most of the planets... human beings made as we [are] could exist'. Gerald Beavis was more sanguine about Mars: 'On the whole it seems agreed that the canals contain water.' Arthur Draper and Marian Lockwood also thought in evolutionary terms: 'The canals might be the handiwork of creatures long dead... [Mars would be] a deserted and lifeless ghost planet upon which we see standing the products of an ancient civilization...'[6]

The mystical thread of pluralism continued, but vestigially. Maurice Maeterlinck's *The Magic of the Stars* (1930) speculated about prehistoric extraterrestrial intervention in Earthly affairs, drawing on hoary analogy and probability: 'it cannot be unreasonable to believe that among the millions of stars... there may be half a dozen on which conditions of life resemble those on our globe.' Like Draper and Lockwood, Maeterlinck wondered if the once-civilised Martians were long perished. The mystic, Sir Francis Younghusband took the matter much further: the Universe might contain matter in forms 'not of life, but of some other quality unknown to us'. In *Life in the Stars* (1926) he suggested that there were five thousand 'possible abodes for life' in a spiritual Universe where 'higher beings... ascended to the level of world-consciousness', governed by 'World Leaders' and indulging their taste for music since 'rhythm is a basic characteristic of the Universe'. Back on Earth, our pluralist yearnings and wonder at the night sky might be occasioned by our sensing the 'harmonious hymns' arising from distant planets.

More prosaically, the 'Martian invasion' affair of Halloween, 1938 when Orson Welles' broadcast of a radio adaptation version of H.G.Wells' *The War of the Worlds* precipitated mass-panic in New Jersey and beyond, demonstrated that although Antoniadi *et al* may have won the technical arguments regarding the Martian canals, pluralism — which arose from the deeps of the human psyche — had most emphatically not been laid to rest (**52**).

At the close of the interwar doldrums there came another harbinger: Sir Harold Spencer Jones (Astronomer Royal, 1933-55) wrote a short, popular book *Life on Other Worlds* (1940). This was an important development in three ways: first, Spencer Jones was an illustrious and widely accomplished astronomer; his prestige therefore greatly aided the restoration of pluralism's credibility. Secondly, the book considered pluralism from a scientist's point of view; it contained little metaphysical speculation. Even where it ventured into probability it was in the new knowledge that the limits of the observable

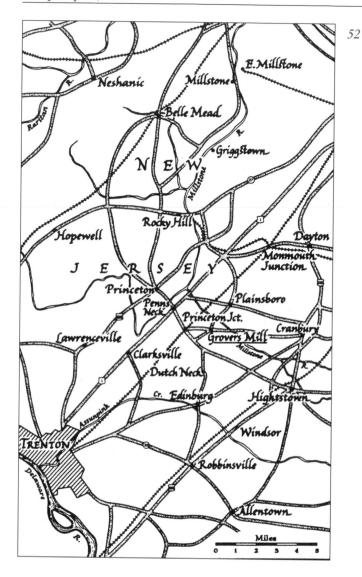

52 *Martian touch-down, NJ: the area in which Orson Welles' panic-generating radio version of* The War of the Worlds *was set, 1938.*

Universe were (then) 500 million light years, and it was set in a careful interrogative: 'Can it be that throughout the vast deeps of space nowhere but on our own little Earth is life to be found?' Thirdly, *Life on Other Worlds* (which enjoyed 11 printings by 1964) was the first of dozens of publications to follow, mainly after the Second World War: books which could draw on the latest science to inform an ever-larger and gradually more scientifically literate public about the new pluralism which, at heart, it had never abandoned entirely.

War and peace

The Second World War effectively put the pluralist debate on ice; it restarted only slowly after 1945. During the war, however, there were scientific and technological developments which later stimulated pluralism. Electronics made possible new types of astronomy by radio, radar, X-ray and other devices, and rocketry provided the basis of post-war space exploration, Earth satellites, and space-based astronomy. Science fiction anticipated some of these ideas and was quick to absorb them; the golden age of the pulps was relatively untarnished by war and lasted into the 1950s. Memories of the Martian canals were fast fading, and prejudices against an assumed Victorian mustiness surrounding pluralism were themselves looking dated, as the sciences enjoyed increasing prestige and technology was able to produce supersonic flight and rocket-based research.

An era which had started unpromisingly with the wreckage of the Martian canal hypothesis ended with a second edition of Spencer Jones' *Life on Other Worlds*; with Arthur C. Clarke's pioneering *Interplanetary Flight* (1950); and for a younger generation, with the adventures of 'Dan Dare, Pilot of the Future'[7] flying to Venus to contend with the malevolence and formidable technology of the 'Treens', green-coloured bipeds who played a part ushering in a new epoch when pluralism, in its scientific and literary forms, was to restore its losses and to advance further than ever before.

References

1   James Jeans, *The Universe Around Us*, 1930.

2   Arthur C. Clarke, *Astounding Days,* 1989; see also Frank H. Winter: *Prelude to the Space Age, the Rocket Societies: 1924-1940.*

3   *Encyclopedia of Science Fiction*, John Clute and Peter Nicholls eds, 1993.

4   Quoted by B. W. Aldiss: *Trillion Year Spree, The History of Science Fiction*, 1986.

5   S. A. Arrhenius, *The Destinies of the Stars*, 1918.

6   James Jeans: *The Stars in their Courses*, 1931; George Forbes: *The Wonder and Glory of the Stars*, 1926; Gerald Beavis: *The Romance of the Heavens*; A. Draper and M. Lockwood: *The Story of Astronomy*, 1940 UK edition.

7   *Eagle* comic, 1950 onwards; 'Dan Dare' was the creation of the gifted artist Frank Hampson.

11 The 'Defining Myth', 1950-2000

Pluralism advanced mightily in the latter half of the twentieth century. By the century's end there existed: a burgeoning pluralist bibliography; entire magazines devoted to its more esoteric manifestations; frequent articles in professional and general journals; films and television series framed around pluralist themes, with lucrative marketing spin-offs: T-shirts, models, toys, computerised games and websites on the Internet.

Pluralism has become well-entrenched, embracing cultural forms from the subject of earnest PhD theses to bizarre cults of alternative lifestyles. Its influence and implications permeate hard science and popular culture in the industrialised world and elsewhere. Karl Guthke suggested that pluralism has become the 'Myth of the Modern Age'.[1] He connected some of the major strands in the pluralist discourse by citing Charles Glicksberg: '[the scientist] must depend on the poetic myth as a means of unifying his vision'.

The causes of pluralism's renaissance and triumph have been threefold: first, the growth of post-war scientific research, particularly in astronomy, cosmology, molecular biology and genetics; secondly, the 'march' of technology: radio and space-based astronomy; improved instrumentation in traditional, optical astronomy and in the life sciences.[2] Thirdly, this science and technology blew on the embers of popular pluralism which had endured through the interwar era. Without exception every advance of pluralism, whether in the sciences or not, has long historical roots. Space engineering, satellites, extrasolar planets, signals to and from space, unusual life forms: no idea has been entirely new under the Sun in late twentieth century pluralism.

The Search for Extraterrestrial Intelligence (SETI)

SETI has been the flagship of contemporary scientific pluralism although relatively lightweight in terms of the resources actually devoted to it. But being directly and unambiguously involved in the search for other worlds, it has captured popular imagination.

The inception of SETI is usually dated from 1959 when two Cornell University physicists, Guiseppe Cocconi and Philip Morrison published a paper in *Nature* 'Searching for Interstellar Communications'. The idea of sending signals to, or receiving them from intelligences in other worlds was of course not new; Tesla and Todd had tried to pick up signals from Mars half a century earlier; the idea of interstellar signalling can be dated from at least Bishop Barnes' suggestion of 1931.

Cocconi and Morrison based their proposition on the possibilities of an actual technology, radio telescopes, which were now up and running. Jodrell Bank telescope in England had started operating in 1957; two years later another started operations at Green Bank, West Virginia; others existed or were about to do so shortly afterwards in Europe, Australia and the USSR.

In 1960 Frank Drake, who had arrived at the same conclusions as Cocconi and Morrison, started a brief programme using the 85ft dish at Green Bank to search for extraterrestrial signals from the vicinity of two stars: Tau Ceti and Epsilon Eridani. He called the enterprise 'Project Ozma'; contemporary pluralism, a child of its time, is a happy hunting ground for codeword and acronym generators and collectors. The 200-hour exercise produced no results, except possibly to demonstrate that pluralism remained controversial. Drake's chief, Otto Struve, was broad-minded about pluralism but had to record soon afterwards that Project Ozma had 'aroused more vitriolic criticisms and more laudatory comments than any other recent astronomical venture...'; the spectre of Lowell still haunted the observatories.

Struve supported the first conference devoted to SETI, held at Green Bank in 1961 and attended by Cocconi, Morrison, Drake, and Carl Sagan who later became SETI's best-known publicist and advocate. Significantly, a biochemist and a zoologist were also present. Sagan later felt of the Green Bank Conference that 'there was such a heady sense in the air... finally we [had] penetrated the ridicule barrier'. As is frequently the case with a new cultural form, SETI explored an astonishing range of possibilities in a short time; its supporters benefited from what J.M.Keynes once referred to as 'exuberant inexperience' and made mileage quickly before orthodoxy, bureaucracy and cost-accountants chilled the enterprise. Also in 1961 J.A.Webb, addressing a conference of radio engineers, suggested listening in for stray extraterrestrial signals; Drake developed this idea, exploring ways in which humanity might detect traces of extraterrestrial life other than deliberate signals, for example the remnants of general broadcasting and communicating. Two years earlier Freeman Dyson had suggested that an advanced extraterrestrial civilisation might dismantle a planet and redistribute it as an energy-gathering shell around a star. This might result in an excessive transmission of infra-red radiation, which could then be detected from Earth. In 1962, the taxonomists George Claus and Bartholomew Nagy prepared for the day when extraterrestrial life-forms might be discovered by offering an extension of the Linnaean classification for outer space.[3]

Soviet SETI was similarly animated. In 1960 Iosif Shklovskii wondered if the two moons of Mars were 'artificial satellites' and suggested the Galaxy might contain 'at least three billion planets... on which a highly organised and possibly intelligent life may take place'. Nikolai Kardashev went further than Dyson in suggesting that advanced civilisations might redistribute energy-gathering matter on a galactic scale. The first Soviet SETI conference was held at Byurakan in the (then) Armenian SSR in 1964; SETI theorists and engineers from the USA and USSR assembled for the first international SETI conference, also at Byurakan in 1971. In 1975 the USSR Academy of Sciences launched a 15-year SETI programme.

The culmination of this freewheeling, creative phase of SETI was 'Project Cyclops' of 1971, at one time studied seriously by NASA, for a mammoth array of radio telescopes

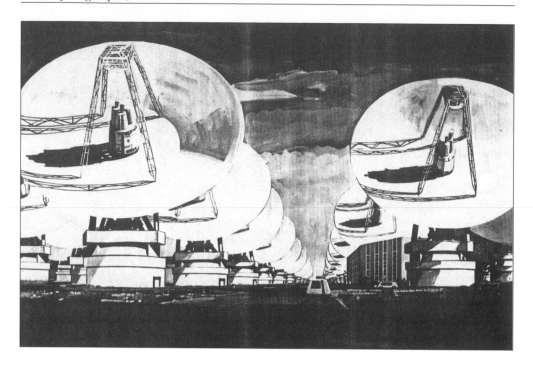

53   An artist's impression of the Project Cyclops proposal; a $10 billion array of 1000 200m dishes for SETI work (1971); NASA preferred more modest schemes; the US Congress still had doubts. (Courtesy NASA)

dedicated to SETI work and costing up to $10 billion over 10-15 years (**53**). The plan did not bear fruit; instead NASA backed the more modest HRMS (High Resolution Microwave Survey), inaugurated in 1992. Even that proved too costly for the US Congress and it was soon terminated; some of its vision and expertise continued as Project Phoenix, privately-financed and buying time on existing radio telescopes.

There have been some fifteen major SETI programmes, and many minor ones, since Project Ozma in 1960, two-thirds of them American but still an international effort; initiatives came also from Argentina, Australia, France, Germany, and the Netherlands, *inter alia*. By 1999 five large projects remained in addition to some amateur investigations, H. Paul Schuch of the SETI League hoped that 'Project Argus' might eventually link 5000 amateur stations The fates of the larger projects indicate the problems: many were financed for a limited duration.

The Soviet programmes faltered after the demise of the USSR. The once longest-running project in the USA, Ohio State University (1973–97) closed down when the land on which its radio telescope was located was sold for development as a golf course; eloquent testimony of cultural priorities.

There have been no unambiguous signals from extraterrestrials in forty years of scrutiny: has pluralism's main practical exercise to date therefore failed to advance its case, even to have undermined it? Given the wearying immensity of the cosmos from which

54 *Signals from Space:
SETI has spawned
some readable novels;
Carl Sagan's*
Contact *became a
popular film; in* The
Listeners *(1972)
messages are received
from a doomed
civilisation which
perishes before its
ultra-long distance
message arrives; a
variation on
'Great Silence'
problems. (Courtesy
Arrow Books)*

signals might come, and the limited scope of SETI programmes, the paucity of results is hardly surprising. In any case 'absence of evidence is not evidence of absence' — the neat maxim of Sir Martin Rees, Astronomer Royal. SETI has also provided the main axis around which the pluralist debate has revolved. It has established the main network within which professional pluralists have carried on their exchanges — conferences, papers, and private discussions, and it provides a public frame of reference (**54**).

Sophists, economists and calculators

The SETI network has become complex; its components such as the SETI Institute are specifically dedicated to pluralist investigation, others are only partly or occasionally dedicated. The IAA (International Academy of Astronautics), IISL (International Institute on Space Law), IAU (International Astronomical Union) and COSPAR (Committee on Space Research) have other main objectives. One outcome of the network's efforts was the 'SETI Declaration' of 1989, adopted by most members within three years.[4] It listed the organisations and authorities that should be approached if a SETI signal were ever to be detected (the list includes the Secretary-General of the United Nations) and how a public announcement might be made. The related 'Dresden Protocol' of 1990, an 'International Policy on a Reply from Earth', stated that a response to a signal from outer space should be sent 'in the name of all humanity' and that 'many issues are not primarily scientific in nature'; a notable advance on the reply to a British parliamentary question of a decade before when it was felt that an extraterrestrial signal ought to be reported to the Post Office, or the BBC.

By the time of the Dresden Protocol there had emerged a small body of theory regarding the legal implications of a SETI discovery or of an actual, physical, contact with extraterrestrial beings. There has also been a debate about which usage to adopt: 'ETs' or 'Extraterrestrials', a neutral term; 'aliens', a tone of defensive hostility; even 'guests', suggested by a Quaker in an attempt to counteract the unfriendly implications of 'alien'.

The diplomacy and jurisprudence of the world was formulated in times innocent of space and ET issues. The first International Space Treaty was signed in 1967 and covered the 'Activities of States in the Exploration and Use of Outer Space, Including the Moon and Other Celestial Bodies'; pluralism's first implicit appearance in the international, or any other, legal framework. But how extraterrestrial visitors should be treated remained an open question. Respected academics had grappled with the problem by the late 1970s: Ernst Fasan in his *Relations with Alien Intelligences* (1970) and Roland Puccetti in *Persons — A Study of Possible Moral Agents in the Universe* (1968). Robert A. Freitas Jnr suggested (1977) that an ET should be classed as an 'essential alien' or an 'alien amy', a 'friendly alien', or if all else fails, a 'pseudoperson' accorded certain rights and responsibilities under 'metalaw'. Puccetti thought that humanity might eventually require 'a new logical term, a "person" able to assimilate conceptual schemes... to experience sensations, emotions... a moral agent'. Without a single extraterrestrial having been discovered, pluralism had already enriched juridical theory.

Popular pluralism

Authoritative popular pluralism was relaunched by Spencer Jones with his *Life on Other Worlds* (1940). Carl Sagan and Iosif Shklovskii produced the fundamental text of the 1960s, *Intelligent Life in the Universe* (1966), a compendium which brought together theory and practice from many sciences, rivalled by Walter Sullivan's popular and rather more accessible *We Are Not Alone* (1964). Both books were catalysts in the rapid spread of public interest in a plurality of worlds, or 'life on other worlds' as the popular phrase put it.

The public's continuing interest in pluralism was sharpened by the achievements of space science. In many books and magazine articles there were implicit or explicit anthropomorphic assumptions about 'life' on other worlds; that it would be in some way organic, oxygen-breathing and water-consuming. These ideas permeated the well-informed *The Green and the Red Planet* by the German physiologist, Hubertus Strughold (1954) dealing with possible life on Mars, and Lord Nelson's *There Is Life on Mars* (1955).

The old pluralist debate about the precise forms which extraterrestrial life might take, an issue well-known to Bishop Wilkins, Christian Wolff or Thomas Dick was renewed in a secular, scientific garb. Carl Sagan, for example, elaborated the concept of 'species chauvinism'. He questioned whether or not life-bearing planets had necessarily to orbit stars like the Sun (G-star chauvinism)[5]; or whether they had to rely on water or oxygen (water and oxygen chauvinisms) etc; with characteristic openness he admitted to having carbon-chauvinistic tendencies himself.[6]

As Americans came to dominate the day-to-day theory and practice of pluralism so American writers came to dominate pluralist literature. The smaller British contribution reflected a relative shrinkage in its influence. The positions within English-speaking pluralism prevalent in the era of Dick, Chalmers, Whewell and Dwight were thereby reversed. One enduring feature remained: the disproportionately high contribution to quality pluralist literature made by Scots, writers like John Macvey *(Whispers From Space*, 1973), Duncan Lunan *(Man and the Stars*, 1974) and Chris Boyce *(Extraterrestrial Encounter*, 1979). English-speaking readers could also draw on another quality tradition of pluralist writing in translation: from the German, of Reinhard Breuer's *Contact with the Stars*, 1982 (*Kontakt mit den Sternen*, 1978) which was unusually comprehensive and readable in a field which tends to veer uneasily between technical density and loose-ranged imagination; and from the French, of Emmanuel Davoust's *The Cosmic Water Hole*, 1991 (*Silence au Point d'Eau*, 1988) and Jean Heidmann's *Extraterrestrial Intelligence*, 1995 (*Intelligences Extra-terrestres*, 1992).

Ancient-astronauts, and UFOs

Two new pluralist genres flourished in the 1970s. Both enjoyed a wide and at times massive readership although they were, in the eyes of most orthodox pluralists, problematic: theories that the Earth had already been visited by astronauts from far-distant worlds, and literature concerned with UFOs (Unidentified Flying Objects). The pioneer theorist of the ancient-astronaut (or 'palaeocontact' or 'palaeovisit') school, Erich von Däniken produced *Chariots of the Gods?* in 1970. The market for such literature seemed at the time to be insatiable; von Däniken produced a further seven books ending with *My Proof of Man's Extraterrestrial Origins* (1977). He had many imitators, the titles of whose works often indicated their scope and focus, for example: Robin Collyns: *Did Spacemen Colonise the Earth?* (1974); Richard Mooney: *Colony Earth* (1974); W. Raymond Drake: *Gods and Spacemen Throughout History* (1975). Even the Moon, long abandoned by most pluralists in fact and fiction as unpromising territory, received its share of treatment in *Our Mysterious Spaceship Moon* (Don Wilson, 1976) and George H. Leonard's *Someone Else is on Our Moon* (1976) **(55)**. Within a decade over 40 million copies of von Däniken's books were printed, a size of market which suggested that this new branch of pluralism answered definite social and psychological needs, like the earlier popular pluralist works of rather greater literary distinction, by Fontenelle, Thomas Dick, or Camille Flammarion.

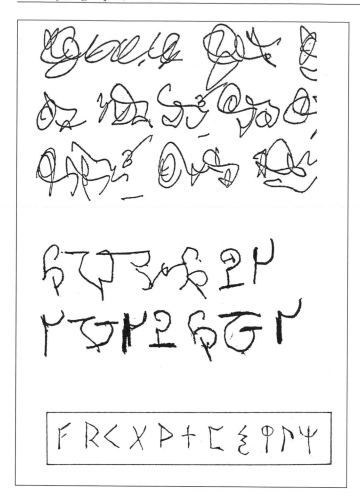

55 *Alien calligraphy: claimed to have been observed by Terrans from top: an extract from* Hamlet *in Venusian (by telepathy, Bernard Byron); Martian writing (Helen Smith, a medium, early twentieth century) and rune-staves from the Moon (George Leonard, 1976).*

In two books, *The Space-Gods Revealed* and *Guardians of the Universe?* (both 1980), Ronald Story demonstrated the methodological weaknesses of von Däniken's works. Chief amongst these was the selection of facts to illustrate a preconceived theory, a tendency not unknown in either orthodox science or history.

The ancient-astronaut school was also fond of rhetorical ploys: 'For all we know there may be... floating around in deep space the remnant testimony of ancient space-wars' — the blurb on the cover of Robin Collyns' *Prehistoric Germ Warfare* (1980). Some of the chapter headings of this book were couched in a suggestive interrogative: 'An Inca City on Mars?'; 'Ultraviolet Lasers from Tau Ceti?' and 'Did a Spaceship Visit New Zealand Six Million Years Ago?', phrases which demonstrate the close connection between this particular branch of literature and pluralist themes.

There was, however, a more measured, scholarly wing to the genre manifest in, for example, *The Spaceships of Ezekiel* (1974) in which J.F. Blumrich, a NASA engineer produced an ingenious interpretation of the Old Testament Book of Ezekiel, Chapter I, or Robert K. Temple's *The Sirius Mystery* (1976), a closely-researched hypothesis about extraterrestrial visitors coming to Earth from a distant star system. The flood of ancient-astronaut literature in the 1970s suggested a strong public yearning for pluralist ideas, not so much of a metaphysical kind but rather of one that seemed to draw on 'hard' science or history, or plausible imitations of both. Pluralism, often of a none too rigorous kind served this hunger for mystery, meaning, and explanation at a time when religions and ideologies seemed to falter.

Books like George C. Andrews' *Extra-Terrestrials Among Us* (1986) linked ancient-astronaut material to the major alternative pluralist theme, UFOs. This subject antedated the ancient-astronaut excitement and it continued after the school of 'gods from space' peaked and quietened in the late 1970s. The UFO phenomenon has a long history; its chroniclers cite examples of strange craft spied aloft in the Middle Ages and there had been a rash of reports about airship-like objects observed in the USA in 1896-97. Modern 'ufology' (the field has developed an extensive argot of its own) dates from 1947. Interest since then has come and gone in waves: 1952; 1965-67; 1973; later 1990s. Much of it has been in the hands of amateurs working individually or in dedicated groups like the Center for UFO Studies in the USA. Official, and professional scientific, interest has been sporadic, mainly from the US Air Force which commissioned the astronomer J. Allen Hynek to produce a report on the subject (1949), then conferred with the physicist Donald Menzel in the 1950s; and once more commissioned an investigation resulting in the *Final Report of the Scientific Study of Unidentified Flying Objects* (1969). The CIA and the US Congress also investigated the phenomenon from time to time. Although interest in UFOs is world-wide, in terms of sightings, investigations and formally-organised networks of observing and reporting, the USA enjoys a marked preponderance, as it does with pluralism as a whole.

There is no necessary connection between UFOs ('flying saucers' to the laity, especially in their early days) and pluralism; nevertheless a connection was soon assumed and it influenced the debate during the height of public interest in the 1960s (**56**). Although explanations of UFOs ranged from natural phenomena to psychological needs (Carl Jung, *Flying Saucers: A Modern Myth of Things Seen in the Sky*, 1959) the favoured explanation was that they came from other worlds in outer space, sometimes nearby (H.J.Wilkins, *Flying Saucers on the Moon*, 1954), sometimes further away. In *Flying Saucers Have Landed* (1953) one of the authors, George Adamski claimed to have encountered a Venusian emerging from a UFO in California — a human form of average height wearing ox-blood coloured shoes and a one-piece suit of 'chocolate brown'.

These books were the harbingers of a literary deluge whose emphasis has changed over time; in the 1980s and 1990s it tended towards UFO crashes, government conspiracies, cover-ups, and the abduction of humans by aliens from other worlds. Its arena has grown *pari passu* with developments in modern cosmology, with theories of UFOs from other dimensions, other universes, other forms of space-time than those prevailing in our corner of the cosmos.

The UFO phenomenon raises important questions of perceptual and social psychology. It has occasionally been suggested that it is a classic myth, even the first new myth of a high-technology world. If so, it is in some measure a myth-within-a-myth, ie within pluralism. It also raises questions of scientific method: scientists and others attending to the issue of UFOs have often argued about the nature of evidence and quality of

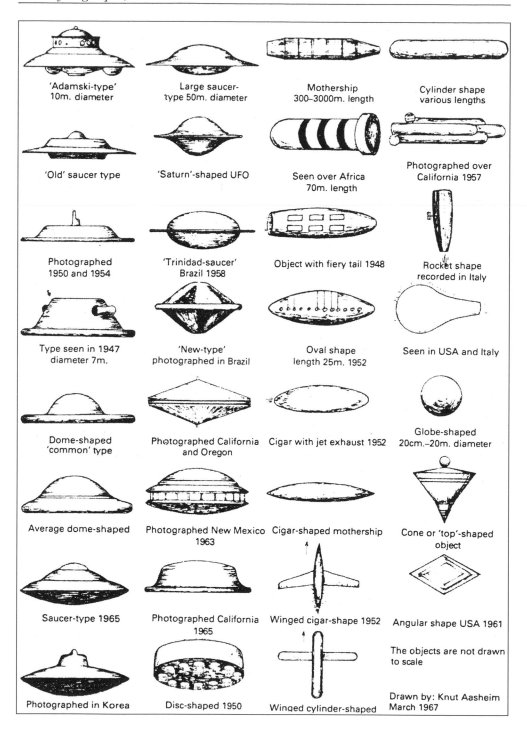

'Adamski-type'
10m. diameter

Large saucer-
type 50m. diameter

Mothership
300–3000m. length

Cylinder shape
various lengths

'Old' saucer type

'Saturn'-shaped UFO

Seen over Africa
70m. length

Photographed over
California 1957

Photographed
1950 and 1954

'Trinidad-saucer'
Brazil 1958

Object with fiery tail 1948

Rocket shape
recorded in Italy

Type seen in 1947
diameter 7m.

'New-type'
photographed in Brazil

Oval shape
length 25m. 1952

Seen in USA and Italy

Dome-shaped
'common' type

Photographed California
and Oregon

Cigar with jet exhaust 1952

Globe-shaped
20cm.–20m. diameter

Average dome-shaped

Photographed New Mexico
1963

Cigar-shaped mothership

Cone or 'top'-shaped
object

Saucer-type 1965

Photographed California
1965

Winged cigar-shape 1952

Angular shape USA 1961

The objects are not drawn
to scale

Photographed in Korea

Disc-shaped 1950

Winged cylinder-shaped

Drawn by: Knut Aasheim
March 1967

56 *Many and various UFOs: Knut Aasheim's 1967 record of forms claimed to have been seen by earth–bound observers.*

investigatory processes as much as the data to hand. An abiding problem with the study of UFOs, as with 'old pluralism', has been that so many of its assertions have been untestable, therefore unfalsifiable. The matter was summarised neatly by the British astronomer and writer on pluralism, Ian Ridpath in his 'UFO Uncertainty Principle': 'one cannot have a UFO sighting which is both highly reliable and highly specific'.[7] Thus, where many people see a UFO it is usually explicable in terms of natural or artificial phenomena; where an extraterrestrial spaceship or being makes its presence known it is seen by few witnesses whose testimony cannot be corroborated.

Science fiction

Science fiction continued to contribute to, and comment upon, the evolution of pluralism. Its best literary products did this effectively, but it also had a long tail of lacklustre work which engaged only marginally with the deeper pluralist questions. Other media took up the baton effectively: films, television, and graphics, for example. 'Space art' experienced a renaissance after 1945, inspired by the paintings of Chesley Bonestell (**57**); frequently depicting images of other worlds, occasionally inhabited, in striking ways. The graphics of science fiction paperback covers sometimes raised as much pluralist consciousness as the texts they illustrated. Films and television frequently embodied time-hallowed pluralist themes and contemporary concerns. Fears of Cold War annihilations and technological catastrophes suffused films like *Rocketship X-M* (1950); *Forbidden Planet* (1956); and the East German *Der Schwiegende Sterne* (*The Silent Star*, 1960). In the latter, explorers from Earth find the remains of an ancient, self-destructed civilisation on Venus. A benevolent alien was the eponymous star of *ET* (1982) but malevolent extraterrestrials also abounded for example in the *Quatermass* films (1955-67) and *Invasion of the Body Snatchers* (1956). Misunderstood aliens appeared in *It Came from Outer Space* (1953); panpsychist pluralism appeared in *Solaris* (1971); ancient-astronauts, implicitly if ambiguously, in *2001 — A Space Odyssey* (1968); and the receipt of a SETI message in *Contact* (1996).

Mainstream science

Although the flowering of UFO and ancient-astronaut ideas excited popular interest in pluralism, the frequently dubious methodology employed may have discouraged serious scientific engagement with the question of life on other worlds. We cannot, of course, know for certain how far scientific pluralism might have advanced if its embarrassing cousins had not irrupted on the scene. We do know that the awarding of research grants, and the consequent career ambitions of promising young scientists, have only occasionally offered direct support to pluralism, for example in some SETI projects. In spite of its huge potential importance, undiluted pluralism can appear a chancy, long-term investment when compared to more urgent issues such as the improvement of weaponry or the alleviation of human diseases and suffering. In spite of this poverty of direct investment, great advances have been made, usually in other fields upon which pluralists could draw or to which they could make a contribution, notably in space programmes.

57   Space Art: an important source of enthusiasm for space science and pluralism; this painting of a 'new' Saturn viewed from one of its moons, was by the doyen of the art form, Chesley Bonestell (1950).

Two areas in particular have enriched pluralism: the life sciences, and astronomy. The striking growth in the study of biology, biochemistry, and genetics has been beneficial to pluralism through the scientific investigation of the origin of life and the evolutionary processes by which it develops. The new science of 'exobiology' (the term dates from the 1970s) is specifically dedicated to studying the biological aspects of pluralism. By throwing the net of biological investigation beyond Earth it has contributed valuably to mainstream biology; the US space agency, NASA, has for example invested heavily in biological research so that astronomers have been able to work fruitfully with exobiologists on the enigma of the origin of life.

NASA interest — and at one time Soviet interest — in exobiology arose partly from the need to study the possible contamination of objects in space which had been visited by missions from Earth, whether manned or not. Once contamination starts it would be difficult, if not impossible, to determine whether or not microbes found on a planet were or were not extraterrestrial in origin. The search for signs of extraterrestrial life became implicit or explicit in most of the American and Soviet space programmes.

By the mid-1990s there had been 35 missions to the Moon, 24 to Venus and 23 to Mars; some flew by or around these bodies, others made landings. The NASA *Viking* expeditions to Mars in 1976, for example, had clear pluralist implications. *Viking 1* took numerous pictures of Mars which immediately excited neo-Lowellite reactions when it appeared that there were some pyramid-like structures and the giant carving of a human face in the Cydonia region of the planet. The 'face on Mars' quickly generated a wave of pluralist literature and conferences. *Viking 2* landed on the planet and took soil samples, duly analysed for traces of organic life. The results were inconclusive, and have fuelled scientific debate ever since.

In 1996 there were two Martian initiatives, one American, one Russian, both suffused with pluralist assumptions. The American *Mars Pathfinder* landed on Independence Day, 1997 promptly disgorging a small vehicle, controlled from far-distant Earth, investigating the soil and rocks of Mars. The Russian *Mars 96*, specifically designed to search for traces of life, contained in its cargo a CD-ROM entitled *Visions of Mars*, bearing over seventy novels, articles and broadcasts chronicling humanity's long fascination with the Red Planet from tenth-century Arabic poetry to contemporary science fiction and documentaries, much of it unashamedly pluralist in tone. The mission failed, however, and the disk and all else lie possibly on the bed of the Pacific Ocean, possibly elsewhere.

Space exploration brought generally discouraging news about Venus. Prior to the expeditions, pluralists created a wide range of Venusian conditions.[8] Donald Menzel and Fred Whipple had suggested in 1955 that the massive quantities of carbon dioxide in the Venusian atmosphere would result in an ocean of soda-water covering the planet; Sir Fred Hoyle thought oxidised hydrocarbons might leave an ocean of oil; G.A.Tikhov in the USSR favoured a great Venusian desert populated with orange-coloured shrubs. C.S.Lewis tried another tack in his science fiction novel *Perelandra* (1943): a Venusian sky 'of pure, flat gold like the background of a medieval picture' under which lay a golden ocean marked by waves which toned down from emerald to a 'lustrous bottle green, deepening to blue'.

Alas for the Venus constructed by the imaginations of scientists and literati. From 1971 to 1981 some 21 space probes (15 Soviet, six American) found that Venus had a surface baking at 464°C, a crushing atmosphere, mainly of nitrogen and carbon dioxide which bore thick clouds of dilute sulphuric acid, struck and illuminated by ferocious bursts of lightning. Undaunted, Carl Sagan suggested sowing Venus' upper atmosphere with algae, and Stephen Gillett outlined a 400-year programme to produce shallow brine-oceans 'suitable only for algae and shrimp' — evidence of another initiative in the pluralist saga, that of 'terraforming': of converting unpromising planets or satellites into habitable worlds on which humanity might one day settle.

Putting existing life on other worlds is, however, a very different matter from finding it already there. In the search for traces of life elsewhere in the Universe, astronomers produced promising leads when they discovered, during the late 1960s, traces of organic material in the molecular clouds of outer space. Within twenty years they could demonstrate that amino acids existed in meteorites, and simple organic matter in comets or in massive interstellar gas clouds, although none of sufficient complexity to suggest that this material was able to produce the necessary conditions for life. Nevertheless, pluralists were encouraged; in 1977 the Russian scientist V.I.Goldanskii felt: 'it seems necessary to consider interstellar grains as a very important stage of prebiotic evolution'.[9]

58   *Mass pluralism: NASA reveals the 'Martian meteorite' bearing possible traces of life, August 1996. (Courtesy NASA)*

During the 1990s science offered other supports and comforts for pluralism: possible congenial habitats for life in the solar system; the discovery of the 'Martian meteorite'; and evidence of extra-solar planets. Spacecraft, flying close to potential worlds hitherto too remote from Earth to be investigated in detail, discovered new candidates for pluralist aspirations. *Voyager 2* (1979), and more particularly *Galileo* (1996) found that Europa, a Jovian moon, was covered with heavily cracked ice which might in turn cover a liquid ocean. At about the same time one 'origin of life on Earth' theory suggested that simple life forms might be generated by deep, lightless submarine vents: if on Earth, why not on Europa? The *Voyager* spacecraft also found evidence of simple organic molecules on Titan, one of Saturn's eighteen moons. Nearer to home, dried-out deltas and ravines implied that Mars had once possessed abundant water; even the Moon — the oldest home of pluralist aspirations — tempted exobiologists and space colonisers with putative stocks of water ice.

In August 1996 NASA announced that a small meteorite found in Antarctica in 1984 had landed there some 13,000 years previously, having originally been struck off Mars by an impact 16 million years before that. NASA phrased its revelations carefully: the meteorite was *possibly* from Mars; it *appeared* to contain the tiny fossil of what *might* have once been a bacterium, a minute living organism (**58**). Although the announcement was cast in tentative language the media were predictably quick off the mark. In Britain within a few days the *Daily Mail* had headlines announcing 'Life on Mars' and a forthcoming 'Armada to Mars'. Sceptics noted that NASA, as a publicly-funded body, needed to generate public interest and support; President Clinton, limbering up for a re-election campaign, also used the meteorite news skilfully. Whatever the truth about this controversial discovery, the fact that it was processed carefully by publicity managers demonstrated the large potential of popular pluralism, at worst dormant, at best rampantly curious, but by now deeply embedded in public consciousness.

For pluralists, perhaps, the most promising scientific discovery was the coming-true of a long-cherished dream (for example, of Chalmers in the nineteenth century and Bentley in the seventeenth century) of finding planets beyond the solar system. The venerable concept of extra-solar planets had appeared in the writings of many pluralists, including Bruno, the Herschels, Chalmers, Kant, and Thomas Dick. In 1995 Swiss then, later, American astronomers presented evidence of extra-solar planets: first one which orbited the star 51 Pegasi; then planets orbiting 47 Ursae Majoris and 70 Virginis; others were discovered shortly afterwards, even traces of an extra-solar systems near Upsilon Andromedae. Although no extra-solar planet had actually been observed, their existence could be inferred from the characteristic behaviour of their parent stars, detected by sophisticated instrumentation and thereby confirming a well-established feature of pluralist history, its close relationship with developing science and technology.

Life everywhere?

One way of recasting pluralist discourse has been to argue that life, particularly in its initial forms, permeates the entire cosmos and that, instead of searching for this or that manifestation of it, we ought to investigate its universality. The leading edge of this argument is 'panspermism', the concept of life suffusing the cosmos; a philosophy with an unusual pedigree. It has tended to exist on the margins of orthodox science although some of its advocates have been amongst the most prestigious of world-class scientists: Lord Kelvin, Svante Arrhenius, Sir Fred Hoyle and Francis Crick.[10]

The panspermists have differed in important details: Kelvin suggested life came to Earth on meteors, 'lithopanspermia'; Arrhenius that bacteria were propelled through space by stellar radiation, 'radiopanspermia'; Crick that primitive life-forms might have been deliberately projected to Earth in some kind of vehicle, 'directed panspermia', an unorthodox and original proposition and therefore one made wisely after he had been awarded a Nobel Prize.

Science fiction has also adopted the idea from time to time; in John Wyndham's *The Midwich Cuckoos* (1957) women in an archetypal English village were impregnated by alien visitors. The imaginative possibilities of panspermism are considerable; Hoyle took full advantage of them in a series of popular scientific books written either alone or jointly with Chandra Wickramasinghe. He at once enriched pluralism and flustered biological orthodoxy which, like any other intellectual enterprise, resents outsiders entering its field. The new panspermism, and its variants, were enshrined in: *Diseases from Space* (1979); *Evolution from Space; Space Travellers, The Bringers of Life* (both 1981) and *The Intelligent Universe* (1983).

Noting that the chances of assembling the full range of proteins upon which life depends as being $10^{40,000}$:1 Hoyle concluded that the likelihood of its appearing

59   *Bringer of life? The Great Comet of 1843 as seen from Blackheath, Kent (now London, SE) — versions of panspermist theory suggest that primitive life-forms are transported about space on comets, evolving further when they land in promising environments.*

spontaneously on Earth were effectively nil, from which premise he speculated that primitive life came to Earth from interstellar space on comets (**59**) or their remnants: 'As ... details about these fossils from space [remnants of organisms purported to have been detected in meteorites] are produced, the evidence of life outside Earth ... begins to fit in place ... if comets were a source of life, of micro-organisms ... in the remote past, so they must be today.' But what assembled these life forms in the first place? 'The intelligence which assembled the enzymes did not itself contain them. This is tantamount to arguing that carbonaceous life was invented by a non-carbonaceous intelligence, which by no means need be God, however.' Thus did one variant of modern, science-based, pluralism connect with the 'big questions' of philosophy and theology which had sustained the old, metaphysical pluralism.

Other partly-metaphysical pluralist ideas can be found lurking in the outer reaches of cosmology. In the 1970s, James Lovelock suggested that the Earth should itself be perceived as a single, living organism of which intelligent humanity was but a component part; the 'Gaia' hypothesis.[11] Lee Smolin of Syracuse University, New York took this concept further: perhaps galaxies were living systems, 'self-organizing cycles of materials and energy'. Taking the matter further still, the Russian A.D.Linde suggested that the

Universe might be part of a self-reproducing system of universes. The implications of evolutionary theory and pluralism which meet and interrelate at this point could not, at the present state of science or metaphysics, go much further in the scope of their speculation. After two and a half thousand years the sentiments of Epicurus in his letter to Herodotus echo hauntingly: 'There are infinite worlds both like and unlike this world of ours ... atoms being infinite in number are borne out into space ... so that nowhere exists an obstacle to an infinite number of worlds.'

References

1   Karl S. Guthke, *Der Mythos der Neuzeit*, trans *The Last Frontier*, 1983.

2   Martin Harwit (*Cosmic Discovery*, 1981) estimated that of the cosmic phenomena known to humanity (from the Moon to gamma ray bursts, etc) about one half were discovered by the science of the twentieth century. Similarly, Zdeněk Kopal (*Widening Horizons*, 1970) estimated that whereas there might have been but fifty astronomers in Galileo's day, the International Astronomical Union had 207 members on its foundation in 1921; 2,800 by 1970. But these figures refer only to professional astronomers; the majority, then as now, were amateurs.

3   G. Claus and B. Nagy, 'Consideration of Extra Terrestrial Taxa', *Taxon*, June 1962.

4   See 'A Declaration of Principles Concerning Activities Following the Detection of Extraterrestrial Intelligence', *Acta Astronautica*, Feb 1990.

5   In the classification employed by astronomers, the Sun is a dwarf star, spectral type G2.

6   Carl Sagan: *The Cosmic Connection*, 1973.

7   Ian Ridpath, *Life Off Earth*, 1983.

8   Attributive names for the planets are generally formed from the Latin genitive, occasionally from Greek prefixes. In the case of Venus this would result in the *double entendre* 'Venerian' or 'Venereal', hence the inelegant 'Venusian' or contrived alternatives, 'Cytherian', and 'Aphroditian'.

9   *Nature* 269, 1977.

10 For a history of panspermism see: H. Kamminga, 'Life From Space' in *Vistas in Astronomy*, 1982.

11 J. E. Lovelock, *Gaia, A New Look at Life on Earth*, 1979.

12 Pluralist enigmas

In spite of the dramatic extension of the theory and practice of pluralism in the previous fifty years, there remained at the century's end no unambiguous evidence of extraterrestrial life. This disappointment for the pluralists could be changed overnight at a stroke. The discovery of the simplest and most basic life-form — a microbe would suffice — if it were clear beyond peradventure that it hailed from anywhere beyond Earth, would be momentous. No gloss would be necessary to shut down a vast, complex and centuries-old field of pluralist and anti-pluralist debate. It would have demonstrated that life on Earth was not unique. Other debates would remain open until higher forms of life, above all intelligent life, had been discovered. Such a discovery would open yet further debate, and so on ... .

Meanwhile humanity waits; the 'Great Silence' reigns — a most eloquent silence if we know how to read the runes.

Anti-pluralism and the Fermi Paradox

The anti-pluralist tradition, still in evidence in the late twentieth century, is a long and not ignoble one; it has kept pluralists on their toes for a long time. Its advocates and supporters have included Plato, Aristotle, St Thomas Aquinas, Melanchthon, Whewell, and Alfred Russel Wallace.

The simple question which has for long haunted pluralists is that, if extraterrestrial beings exist, why is there no hard evidence of them? This conundrum lay behind Whewell's sobering-up of pluralist debate in the 1850s. A century later it was reiterated by the Nobel Laureate and physicist, Enrico Fermi. In 1950 he asked of some fellow nuclear physicists, apparently at a summer-time lunch given at the Fuller Lodge, Los Alamos: 'If there are extraterrestrials, where are they?' This question, obvious to the point of banality, has been invested with the dignity of a formal title, the *Fermi Paradox*, reflecting its prestigous source. Because he was fond of pointing to the case *for* pluralism, Fermi's question was paradoxical in its setting although, since it is usually taken out of context it hardly appears to be paradoxical in form. Irony surrounds the birth of the Fermi paradox, apparently lost on most commentators: Fermi and his colleagues were engaged on research leading to the production of 'weapons of mass destruction'. Such scientific endeavour, replicated elsewhere in the cosmos, might offer an explanation for the Great Silence, for the non-arrival of extraterrestrial spaceships or messages.

From the mid-1970s the onward and upward rush of postwar pluralism met the first of a series of checks. All to the good: any proposition can benefit from the toughening it gets in battle. Pluralism has not been immune from the rule that unconstrained enthusiasm can lead to facile optimism, exaggerated claims, uncritical flights of the imagination and, in turn, to diminishing returns of interest and commitment from a public that, whatever its sympathies, becomes bored with even the most engaging subjects. The late-twentieth century negative reaction to pluralism can be dated fairly precisely.

In 1975, in the USSR, the leading pluralist Shklovskii, concerned by the Fermi Paradox — the 'Astrophysical Paradox' in Soviet usage — 'defected' to the anti-pluralist camp, announcing his change of mind at an All-Union conference on CETI (Communication with Extraterrestrial Intelligence). He wrote later: 'An analysis of the facts precludes, with a high degree of probability, the possibility of super-civilisations existing, not only in our Galaxy but also in all the local systems of galaxies.' He went on to suggest that at least extraterrestrial civilisations which had developed beyond the primitive level 'must take the road of unlimited expansion'. He thought 'the activity of... a super-civilisation could not have escaped the attention of Earth–based astronomers' but there was no evidence of this cosmic imperialism; 'you cannot secrete an awl in a bag', he concluded, employing one of the cryptic proverbs dear to Russians.[1]

Similar arguments appeared in two papers published in the West in the same year. Michael Hart of the National Center for Atmospheric Research in the USA wrote 'An Explanation for the Absence of Extraterrestrials on Earth'[2], in which he noted: 'There are no intelligent beings from outer space on Earth now.' Hart explored the feasibility of space-travel; the leakiness of some theories which suggested extraterrestrials might have no desire to explore the Universe; the cosmic time available for explorations to occur; and the possibility of ancient-astronaut visitations. He found most pluralist ideas about inhabitable worlds, the evolution of non-human civilisations and SETI contact 'very exciting prospects indeed' but then he administered a cold douche: 'wishful thinking may lead us to overestimate the chances that the conjecture is correct'. His 'Fact A' (that extraterrestrials are not here) led him to two conclusions: SETI was 'probably a waste of time and money' and 'in the long run, cultures descended from ours will probably occupy most of the habitable planets in our Galaxy'. This last proposition, of 'terraforming pluralism', already inherent in the concept of space-travel has developed in parallel with the traditional forms of pluralism. Writing for the British Interplanetary Society, David Viewing argued along the same lines: 'All our logic, all our anti-isocentrism argues that we are not unique — that they must be there. And yet we do not see them.'[3]

The colonisation controversy

These two papers, and the energetic reaction to them, indicated that pluralism was entering one of the periodic crises from which it has generally emerged leaner and stronger. There was a flurry of papers about the feasibility of interstellar colonisation and the possibility of space-probes sent out by extraterrestrials. In 1979 sceptics and pluralists met to consider the Fermi Paradox at a conference in Maryland. They touched on a wide range of issues: SETI; UFOs; long-range space-travel; space colonies; and extraterrestrials lurking in the asteroid belt. No dramatic conversions or conclusions emerged, except perhaps that pluralism was entering another of its problematic phases.

A further blow to pluralism followed in 1981. Frank Tipler, a mathematical physicist from the University of Texas emerged as a leading anti-pluralist with his uncompromisingly entitled paper; 'Extraterrestrial Intelligent Beings Do Not Exist'.[4] Building on Hart's arguments, Tipler argued that any advanced civilisation in outer space would have developed 'a self-replicating universal constructor', a form of robotic space probe, the so-called 'von Neumann machines' (named after their putative inventor), able to repair and reproduce themselves, so that their myriads could have explored, colonised or investigated the Galaxy more effectively than sending out radio signals.

Tipler reinforced his assertions with calculations purporting to demonstrate that only one intelligent civilisation, Earth-bound humanity, had emerged in an assumed 15-billion year history of the Universe. He also employed an unusual and effective stratagem; comparative history. In his 'Brief History of the Extraterrestrial Intelligence Concept'[5] he noted that pluralists had historically depended on the principle of plenitude; on the assumption that an infinite Universe would contain an infinity of worlds; and on a certain innocence of the cultural setting of their own beliefs. By quoting Fontenelle he was able to show that the Fermi Paradox and latter-day scepticism had a long ancestry: 'The people of the Moon would therefore have come to us before now, replied the Marchioness [Marquise], almost in anger.' To the philosopher's specious reply that the Moon-people will come to Earth 'when they shall be more able and more experienced' the Marchioness — implicitly speaking for the Fermist sceptics of the 1970s — responded: 'You are insupportable... to push things so far with weak reasoning.'

Tipler also drew attention to the similarity between Carl Sagan's invocation of the 'principle' or 'assumption' of mediocrity [that Earth is not special, it is a typical planet orbiting a typical star in a typical galaxy] and William Derham's analogy that because there is a 'Parity and Constant Uniformity observable in all God's works, we have every reason to conclude that every Fixt Star hath a systeme of planets...' He cited Whewell and Wallace with approval, adding in respect of the latter: 'the great evolutionists have always been united against ETI (Extraterrestrial Intelligence)'.

Finally, Tipler suggested that recent advances in science and technology had tended to favour pluralism: the return to fashion in astronomical discourse of the nebular hypothesis regarding the formation of the solar system; biological experiments relating to the Haldane-Oparin hypothesis; the development of radio astronomy which encouraged SETI projects; and, in the 1950s at least, the steady-state theory of cosmology, which argued for a Universe infinite in space as well as time. He concluded: 'I contend that, as has been the case for 2000 years, these philosophical and theological beliefs are the main motivations for a belief in extraterrestrial intelligence.' In short, pluralism was at heart a metaphysical construct, its more recent donning of scientific apparel notwithstanding.

Inviting trouble?

Anti-pluralism came in other forms. When he learned that the arch-pluralists Sagan and Drake had encouraged the transmission of an interstellar signal from the Arecibo Radio Telescope in 1974, Sir Martin Ryle, Nobel Laureate and Astronomer Royal, reacted strongly, asking the International Astronomical Union to discourage all such further attempts at advertising the presence of intelligence on Earth to potentially hostile recipients elsewhere in the Universe (**60**). His critics noted wryly that shorter-wave transmissions, like television broadcasting had been made from Earth for some forty years. Neither Ryle nor his opponents mentioned that there might be difficulties for any intelligence wishing to process the Arecibo transmission, if it lacked the blessings of high mathematical skills and possession of the patterns of apprehension and perception enjoyed by *homo sapiens* on planet Earth.

Another astronomer with doubts about applied pluralism was Zdeněk Kopal, of Manchester University. He thought of extraterrestrials: 'We might find ourselves in their test tubes' and of SETI he warned: 'if the space phone rings, for God's sake let us not answer it'. These and similar sentiments such as W.H.McNeill's 'Contact between men has shown that those who have the power use it' were often juxtaposed deliberately with the 'contact optimism' of scientists, like Drake and Sagan, compounded of notions linked by contingence rather than logical necessity. For example, Drake had opined in 1976: 'Interstellar contact would undoubtedly enrich our civilisation with scientific and technical information... it is extremely likely that any civilisation we detect would be more advanced than ours... it would provide a glimpse of what our own future could be... we could understand the way of life most likely to be best for us ...' — a set of propositions which owes more to the ingenuous decencies and assumptions of old liberalism than to natural science or historically informed *realpolitik*.

The contact-optimists ran into opposition because of the rivalries inherent in the scientific community and also because their relentless optimism grated on more cautious spirits. There were also more serious issues at stake. In spite of their being sent 'on behalf of humanity', the Arecibo signal and the messages to extraterrestrials inscribed on metal plaques affixed to the *Pioneer* (1972-73) and *Voyager* (1977) spacecraft, were not the outcome of any real, systematic consultation of 'humanity' or its representatives (**61**). The gold-plated long-playing record attached to *Voyager*, still wending its way beyond the solar system, bears sentiments of goodwill and reproductions of a wide range of Earthly sounds and achievements including a Bach concerto; Louis Armstrong's *Melancholy Blues*; and a crying baby. But any half-competent extraterrestrial able to check these politically-correct and exquisitely-chosen messages against other human accomplishments — the Somme; the Holocaust; atrocities in every continent; the relentless abuse of humans and animals; might conclude that the Terrans spoke with forked tongues, hypocritical at best and treacherous at worst.

The Drake Equation

In their regrouping after the 'galactic colonization flap'[6] the pluralists moved more cautiously. The loss of Shklovskii's support seriously discouraged pluralism in the USSR which thenceforth slipped away behind American-dominated efforts. The need for hard evidence stepped up the number of SETI exercises from 15 in 1961–75 to 40 in 1975–95, although all were dependent on the fickleness of short-term funding; they were not the components of a consistent, thought-out global SETI strategy. Pluralists continued to study and contribute to initiatives in astronomy, space science and biology. The

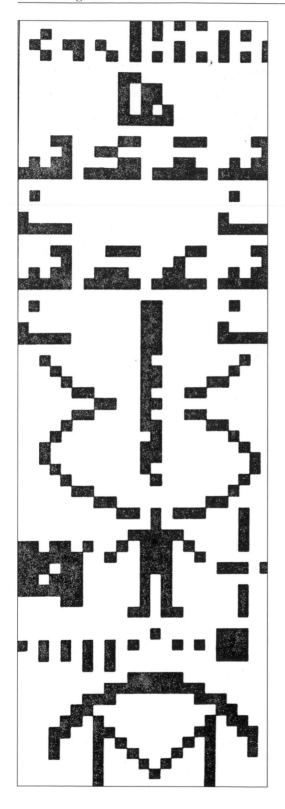

60   The Arecibo message: the visual
      decryption of 1,679 characters
      transmitted in binary code showing
      binary 1-10 (at top), a DNA
      molecule, a human form and the
      Arecibo radio telescope dish; simple to
      process for any being with higher
      mathematics and the typical gestalt
      perceptions of Earth-bound humans.
      The message was beamed at the M13
      star cluster, 25,000 light years away
      — a minimum discourse relay of some
      50,000 years; a triumph of optimism.

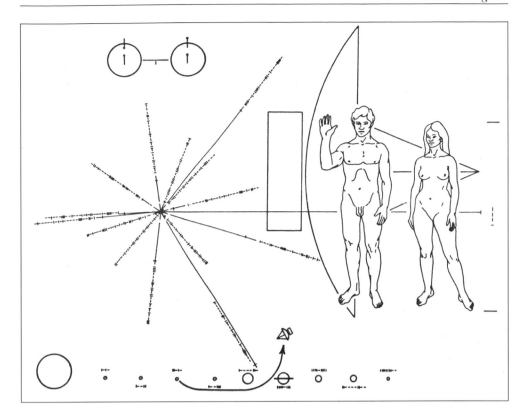

61   *Another message: the 9-inch gold anodized plaque attached to the US Pioneer 10 and 11 spacecraft, launched 1972 and now heading out of the solar system (represented diagrammatically at the bottom edge); other images include a human couple containing 'characteristics of all human races' and the location of the Sun via-à-vis 14 pulsars and number 8 in binary form; extraterrestrials can make of it what they will.*

fundamental causes and effects of pluralism remained. Alvin Toffler put it thus: 'What we think, imagine or dream of cultures beyond Earth not only reflects our hidden fears and wishes, but alters them.'

The fundamental question: 'is there life out there?' requires not only evidence but also a framework within which to situate it and from which to pose appropriate questions. At the time of the Green Bank Conference (1961) Frank Drake produced a simple formula, into which data could be fitted, as and when it was to hand, so as to determine how many extraterrestrial civilisations might actually exist. The *Drake Equation* is set out thus:

$$N = R^\star \, fp \; ne \; fl \; fi \, fc \; L$$

'N' is the number of 'observable civilisations' in the Milky Way Galaxy, the outcome of multiplying the variables to its right. The first few variables are becoming better known owing to advances in astronomy. Thus $R^\star$ is the 'rate at which stars are born' in our

131

Galaxy, now believed to be on average one per year. Next, fp is the 'fraction of stars with planets'; the variable ne is the 'average number of Earth-like planets'; fl is the 'fraction on which life forms'; fi the 'fraction of those seats of life where biological evolution produces intelligence'; and fc is the 'fraction of intelligent civilisations capable of interstellar transmission by radio'. 'L' is the joker of the pack; the 'average lifetime of a communicating civilisation.'

As we move along the set of variables the questions which they pose become ever more difficult to resolve. The discovery of planets around distant stars during the 1990s suggests that fp may be about 5%, or 0.05. On the other hand, there is no available evidence regarding the number of possible life-bearing planets; pluralist optimists hope that future investigations on Mars and Europa will reveal further examples, but once we pass the variable ne we enter the realms of conjecture. It may be that fi is higher than was once thought given the abundance of amino acids throughout the Universe, but as for the emergence of intelligence -who can say? The Harvard biologist Ernst Mayr commented that physicists are apt to be more deterministic than biologists: 'they tend to say that if life originated somewhere, it will also develop intelligence in due time. The biologist... is impressed by the improbability of such a development', a line of argument clearly consistent with Alfred Russel Wallace's stance on the singularity of life on Earth.

Late twentieth-century astronomical discoveries generally heartened the pluralists although their hopes have been overshadowed by the simultaneous revival of catastrophism in the sciences. The revival included studies of: possible celestial collisions between the Earth and asteroids, comets or giant meteorites, of a kind which might have been the cause of the extinction of the dinosaurs 65 million years ago; the instability of the Earth's climate which has been unusually kind for the ten thousand or so years within which our present civilisation has grown; vulcanism and the movement of tectonic plates; and the fortuitous location of the massive planet Jupiter which, by sweeping up cosmic debris, may have saved the Earth from numerous lethal, meteoric bombardments. A change in any one of these conditions might have doomed intelligent life or simply removed the chances of its emergence; if in the solar system, why not elsewhere?

Estimating the probable number of extraterrestrial civilisations antedates the Drake equation, which is but an economical and elegant expression of the obvious. It has been criticised for multiplying factors unnecessarily, for omitting important factors, and for being no more than 'a way of compressing a large amount of ignorance into a small space', the view of the astronomer, Bernard Oliver. Until all components of the equation can be quantified with good evidence it can tell us little; estimates tend to reflect the standpoint of the estimator, pluralist, anti-pluralist, optimist, or pessimist. Frank Tipler concluded that there was one galactic civilisation, that on planet Earth. Other estimates include an assumption of 'one superior community per 1,000 likely stars' (Ronald Bracewell *The Galactic Club*, 1974); 'the number of planets in our Galaxy on which a technological civilisation is now in being = 530,000' (Isaac Asimov, *Extraterrestrial Civilisations*, 1979); and 'the number of extant civilisations substantially in advance of our own in the Galaxy today appears to be perhaps between 50 thousand and one million...'(Shklovskii and Sagan, *Intelligent Life in the Universe*, 1966).

The Great Silence

Whether there is one or more advanced civilisation in our Galaxy or 530,000 is open to debate; the absence of unambiguous messages or visitations from extraterrestrial civilisations (*pace* ancient-astronauts and UFOs) is at least one hard fact. This absence of specific contact was named *The Great Silence* by the American science fiction author David Brin who analysed possible reasons for the silence;[7] other commentators have added their own.

Speculation concerning the enigmatic 'L' of the Drake equation can offer sobering thoughts for the futurologists and social analysts of our own planet. First, there are unfavourable astronomical and biological conditions: inhabitable planets may be rare; life and intelligence may be unusual phenomena; an interest in technology may be uncommon amongst intelligent beings; higher civilisations may be averse to taking risks or publicising their existence; they may even dwell on 'mirror worlds' invisible to us, and vice-versa. Secondly, it may be that our existence is known to extraterrestrials but they prefer to make no contact: we may be inferior beings kept, as it were, in a zoological garden or wildlife park (the 'zoo hypothesis'); or, they await our emergence from barbarism; or, the Earth is in some way unattractive; or, as some pluralists argued at the time of the space colonisation controversy, interstellar flight may be cost-ineffective. Thirdly, the Earth may have been visited or contacted, either in the past or currently, but we cannot detect evidence of such visits. But fourthly, and ominously, perhaps high-technology civilisations do not last for more than a few thousand years in any case.

This last category resonates with contemporary neo-catastrophism and with a host of cyclical theories regarding the rise and fall of civilisations. The possibilities include: cosmic disasters such as asteroid impacts, gamma-ray bursts, failing stars, unfavourable climatic shifts; outbreaks of new diseases; human-induced disasters such as overpopulation; resource-blowing; pollution; violence allied to high-technology weaponry; an 'intelligence ceiling' beyond which science and technology cannot develop; moral dissolution; a failure of will; combinations of these failings; Faustian curiosity and ambition satisfied at the price of an ecological catastrophe.

A moral cosmos

When contemplating the Universe, the Rev Dr Charles Pritchard, (62) Savilian Professor of Astronomy at Oxford 1870-1893, thought: 'surely this steady, various march of the material cosmos can hardly fail to be a type of the moral universe circling round the centre of infinite perfection', a passage which might have earned the approval of Thomas Wright or Immanuel Kant but would probably perplex the typical post-modern mind. The strong infusion of moral questions in the religious and metaphysical earlier pluralism was diluted after the mid-nineteenth century change of emphasis towards scientific pluralism. In a sub-text of his classic *The War of the Worlds* (1898), H.G.Wells has his narrator composing a paper 'on the probable development of morals in the next two hundred years', at the very time the Martians are landing on Earth with the intention of subduing it. At the close of his narrative he returns to his house to find the essay still on his desk, with its implicit

62 *Rev Dr Charles Pritchard, Savilian Professor of Astronomy at Oxford 1870-93; advocate of a mildly evolutionary pluralism but with occasional Whewellite, anthropocentric leanings: 'Who knows? They [the planets] may be just chips struck from the block in the making of our Earth!'*

questions now answered; the ruthlessly Darwinist morality of the Martians suggests what kind of 'moral development' may lie ahead.

Contact pessimists may have some analogic considerations in their favour. *If* the principle of mediocrity holds throughout the Universe, and *if* there is evolution by natural selection, and *if* the behaviour of *homo sapiens* towards other species is typical of those at the head of a chain of predation, *then* the putative existence of more evolved extraterrestrials suggests that we should 'be afraid; be very afraid'.

The moral economy of possible extraterrestrials is not a matter of mere theoretical or passing interest. Questions about their values, motives, ethical systems, and whether or not any of these threaten humanity or are transferable to people on Earth, might be of crucial importance. But this type of question has been generally swept into corners by the secular temper of postwar pluralism.

One of the 'big questions' implicit in pluralism is 'How should we behave towards extraterrestrial beings?' This has produced rational, but largely instrumental, procedural, and juridical discussions; thoughts about 'xenopsychology', 'metalaw' and 'pseudopersons' for example. Current debates about the moral status of animals may

63   *Thomas Carlyle (1795-1881), of the stars: 'A sad spectacle. If they be inhabited what a scope for misery and folly. If they be not inhabited, what a waste of space.'*

throw some incidental light on this vexed and under-discussed field, and even benefit from it. Roland Puccetti considered, for example, that an alien could be accorded the status of 'person' if demonstrably a 'moral agent'. On the other hand, an American lawyer Andrew D. Haley suggested in 1956 that a confrontation with extraterrestrial beings of a culture very different from that of humanity might make it necessary to adjust Matthew VII, 12 ('Therefore all things whatsoever ye would that men should do to you, do ye even so to them') to: 'Do unto others as they would have you do unto them' — moral relativism without even a sliding scale of norms.

Theologians have continued to show a sporadic interest in pluralism but often by recycling existing ideas; lacking the richness and subtlety offered by Derham, Chalmers, Whewell or Pohle, or the medieval scholars. Also, their interest has not been to explore the moral economy of pluralism so much as to reconcile the concept of extraterrestrial life with scripture and the uniqueness, or otherwise, of the Incarnation.

Discussion by such eminent pluralists as Huygens, Gassendi, Berkeley and Chalmers, about the moral nature of extraterrestrials was prominent in the era of metaphysical pluralism. To some commentators, like the Baptist minister Andrew Fuller, extraterrestrials were free from sin, living in a state of grace (*The Gospel its Own Witness*, 1799); others thought that some extraterrestrials might be sinless, but others imagined that they might be corrupted like humanity on Earth. Andreas Ehrenberg, writing in 1710, felt 'it is very possible that in one world the rational creatures have remained in their state... of innocence, while in another they too have been led astray...' To Thomas Wright and Immanuel Kant there was a scale of the moral excellence of inhabitants depending upon how far a planet was from the Sun. Kant wondered if the Jovians might not be 'too noble and wise' to sin. There were many variations on these themes: planets inhabited by angels; planets for purging the souls of the fallen; planets as staging-posts to moral perfection. Karl Gauss, a convinced pluralist, thought that after death our souls inhabited other worlds, including the Sun, a view which converged with mystical pluralism.

Modern scientific pluralism takes little cognisance of the issue of extraterrestrial morality, however. The possible denizens of other worlds tend to be valued for the techniques they might be able to teach us: 'I expect an alien civilisation to bequeath us vast libraries of useful information, to do with as we wish. This "Encyclopedia Galactica" will create the potential for improvements in our lives that we cannot predict' and later: 'I suspect that immortality may be quite common among extraterrestrials...' (both from Frank Drake and Dava Sobell, *Is Anyone Out There?* 1991). The same authors speculate that 'a civilisation of immortals' would be active in SETI work partly 'to extend their resources for amusement beyond their own planetary system.'

In asking what an extraterrestrial civilisation might 'want of us' Sagan and Shklovskii thought that an advanced society would probably not need humans as slaves or as food. They then raised a few desultory questions about 'extraterrestrial evangelism', or the crushing of humanity by extraterrestrials with 'darker motives'; but dark to whom, and by what criteria? They raised also the question that humanity may not yet be 'ready' to join the Galactic community. The terms of this readiness are unclear, although implicitly they are concerned with the need to contain the relentless violence that has cast its pall over human history, a sentiment steeped in the post-Enlightenment liberal assumptions that characterise most contact-optimism.

Some psychologists and anthropologists[8] have offered useful, practical advice on the 'contact question'. The long and venerable tradition of moral pluralism is, however, mainly kept alive and well in fictional form; space operas, for example, have often grappled with the moral implications of intruding on the customs, practices, and even the living space of extraterrestrial beings.

During the Great Moon Hoax of 1836 the ladies of Springfield, Massachusetts raised funds to support a missionary enterprise to the lunarians. In postwar pluralism the notion that humanity might have belief systems or moral codes worthy of exporting to extraterrestrials hardly arose; arguably evidence of a loss of cultural momentum and the spread of post-modern relativism.

Nevertheless, the question was addressed by the 'Oxford School' of pluralists. The Professor of Mathematics at Oxford University, E.A.Milne, a former student of E.W.Barnes, the Bishop of Birmingham who had suggested the possibility of interstellar radio transmissions (1931), speculated in 1950 that developments in radio astronomy might make it possible to beam knowledge of the Incarnation into outer space. The Oxford theologian E.L.Mascall doubted, however, that mere knowledge of the Incarnation was sufficient grounds for 'universal salvation' although he granted that there were no theological reasons for denying redemption to extraterrestrial 'rational, corporeal beings who have sinned...'.

C.S.Lewis, at that time a Fellow of Magdalen College, also took a close interest in these questions, some of which were treated readably in his 'cosmic trilogy' of science fiction classics where the struggle between good and evil was re-enacted on Mars and Venus. Lewis also wrote a monograph in 1959 *Shall We Lose God in Outer Space?* a neat latter-day summary of the 'extraterrestrial morality debate', considered from a Christian point of view, which rehearsed many of the age-old theological problems presented by the concept of extraterrestrial life. In the course of this essay and elsewhere he wondered, 'Whether the vast astronomical distances may not be God's quarantine regulations' preserving extraterrestrials from 'the spiritual infection of a fallen species'.

A related argument, although not set in Christian terms, was presented by Sir Fred Hoyle when contemplating the vast distances to be traversed by any species seeking contact with another over the tracts of space: 'colonization of the Galaxy is impossible because it was deliberately arranged to be so'.[9] Pluralism has often connected with great cosmological questions: the extraterrestrial evolutionism of Proctor; the moral hierarchies in the galaxies of Wright and Kant; the notion of the Universe as an experiment. The Enlightenment scientist Georg Lichtenberg asked: 'Why should there not be stages of spirits reaching up to God; why should not our Universe be a mere experiment, the work of prentice hands?' — an echo of the 'Russian battleship' theory of the Universe: brilliantly conceived, poorly designed, atrociously constructed.[10]

Approaching the matter from the standpoint of another faith, Rabbi Norman Lamm offered a sane set of ideas in his 'Jewish Exotheology'.[11] He argued that pluralism is compatible with Judaism which has a long connection with it. For example, he cited the eleventh century authority Rabbi Judah ben Barzilai of Barcelona who thought there might be 18,000 other worlds bearing intelligent beings. But Lamm declined to become excited about the prospect; looking at SETI and the 'space colonization' excitement he concluded that 'it is earlier than they think' and that the technical demands and costs of space travel were far higher than enthusiasm-dazzled experts supposed. Lamm also warned uncritical pluralists of the dangers inherent in unrestrained pluralism, of the implicit demotion of humanity: 'never before have so many been enthusiastic about being so trivial'.

The relative weakness of religion *vis-à-vis* science as organiser and informant of the pluralist debate may have contributed to its moral impoverishment in modern times quite as much as secular liberalism, moral relativism and post-modernism. Had mainstream pluralism and its chief technical initiatives been set in an enduring Marxist-Leninist USSR or a fundamentalist Islamic culture then its tone, and the style of messages projected into space, might have been entirely different. But that is counterfactual pluralist history; the record demonstrates that the secular West, and the liberal-scientific culture of the USA in particular dominated the practical pluralism of the late twentieth century.[12]

Aspects of pluralism have nevertheless remained on the theological agenda, even if not conspicuously enriched by new thinking. A spate of articles and papers in the 1950s demonstrated a convergence between Catholic and Protestant theologians; Father Grasso writing in *Civietà Cattolica* thought that the concept of extraterrestrials was neither supported nor denied by theology (a view expressed by Brewster a century earlier, and by Campanella and Wilkins over two centuiries before that) — Catholics were free to chose either way 'the last word lies not with theology ... but with the empirical sciences'.

Professor Redeker, a protestant theologian of Kiel University, advanced a similar view: 'Nothing in the Christian faith excludes the possibility of God's creative will select[ing] other heavenly bodies as the habitat for other mortals...' Forty years later, Michael Ashkenazi interviewed 21 religious authorities and concluded that neither the 'Adamist religions' (Judaism, Christianity and Islam) nor the non-Adamist faiths like Buddhism or Taoism experienced any great difficulty in handling the concept of extraterrestrial intelligences.[13] Recent theological writing, however, demonstrates creative thinking on the Incarnation issue. Father Kenneth Delano, writing *nihil obstat* suggested: 'Any one or all three Divine persons in the Holy Trinity may have chosen to become incarnated on one or more of the inhabited worlds of the Universe.'

Anthropocentrism once more

The anthropocentric cast of mind which caused medieval theologians and many of their successors to resist pluralism has returned in secular guise at the end of the twentieth century. The Anthropic Principle suggests broadly that the Universe is arranged as it is, so as to produce intelligent life; any minor deviation in its arrangements would have failed to produce this end. In its 'weak' form it has merely stated 'what we can expect to observe must be restricted by the conditions necessary for our presence as observers' but it has 'stronger' versions which include 'the Universe must be such as to admit the creation of observers at some stage'. Frank Tipler, scourge of the pluralists in the 1970s, concluded that the Universe had indeed brought about intelligent life, but only once, on Earth: 'only on that unique planet on which it occurs is it possible to wonder about the likelihood of intelligent life'. The anthropic principle has run into heavy opposition; to Heinz Pagels it was needless conceptual clutter, to Sir Fred Hoyle it had the 'position inverted... it is not so much that the Universe must be consistent with us as that we must be consistent with the Universe'.

Some of the 'strong' anthropists concluded that the role of Earth-bound humanity was to sally forth, colonise and even reorganise the physical universe. This view gives the 'plurality of worlds' a singular twist, of a kind which would have doubtless excited the condemnation of C.S.Lewis and the approval of Alexander the Great who was reported by Plutarch to have said of other worlds: 'Do you not think it a matter of lamentation that, when there is such a vast multitude of them, we have not yet conquered one?'

In 1845 the Rev George Gilfillan, a Scottish pluralist commented: 'May not this

Universe be only beginning to be peopled, and the Earth be the first spot selected for the great colonization.' These observations demonstrate the long roots of contemporary pluralist thoughts, and the difficulty of claiming absolute novelty for many pluralist ideas.

Is this terraforming pluralism the most noble of pluralist speculations or the ultimate anthropocentric hubris? Two theses may throw light on the issue: first, that the technical ingenuity of space engineering proposals varies inversely as to their economic and political realism; secondly, that the greatest unexplored space in the Universe lies under our own skulls. Thomas Carlyle (**63**) sensed the escapist element in pluralism in 1850: 'Whirl from place to place, at the rate of fifty, or if you like of five hundred miles an hour: you cannot escape from that inexorable all-encircling ocean-moan of ennui. No, if you would mount the stars, and do yacht-voyages under the belts of Jupiter, or stalk deer on the rings of Saturn, it would still begirdle you...'[14]

Contemporary pluralism, like its predecessors, raises profound questions about origins and purposes, of the cosmos, of humanity. Like the pluralism of the Greeks or the Victorians, or anyone else, it expresses the aspirations, fears and conceits of its day. Two of its contemporary features exemplify this well: astonishing technical virtuosity, and moral confusion. It is therefore no monolith, but a scattering of propositions, some widely shared, others minority views, and a remarkable proportion displaying continuities with their historic past.

References

1 *Yearbook of Astronomy*, ed Patrick Moore, 1980.
2 *Quarterly Journal of the Royal Astronomical Society*, 16, 1975.
3 'Extra-Terrestrial Technological Communities' *Journal of the British Interplanetary Society*, 28, 1975.
4 *Quarterly Journal of the Royal Astronomical Society*, 21, 1981.
5 *Quarterly Journal of the Royal Astronomical Society*, 22, 1981.
6 The words are those of Steven J. Dick: *The Biological Universe*, 1996.
7 Quarterly Journal of the RAS, 24, 1983; *Analog Science Fiction/Science Fact*, July 1985.
8 See Barbara Moscovitz 'The Moral Obligations of Anthropology' in M. Maruyama and A. Harkins, Eds, *Cultures Beyond Earth*.
9 F. Hoyle: *The Intelligent Universe*, 1983.
10 An adaptation from the American science fiction writer and naval historian, Fletcher Pratt (1897-1956); cf Lord Francis Jeffrey: 'Damn the Solar System. Bad light; planets too distant; pestered with comets; feeble contrivance; could make a better myself.'
11 'Religious Implications of Extraterrestrial Life' *Faith and Doubt - Studies in traditional Jewish Thought*, 1971.
12 But see Paul Davies: *Are We Alone?*, 1995, especially Ch 6, 'Aliens and Religious Consciousness' which pays some attention to this issue.
13 *Space Policy*, 8, 1992.
14 'Jesuitism' VIII in *Latter Day Pamphlets*.

13 Pluralism, past and present

In the preface to *Aleriel, or a Visitor From Other Worlds* (1883) the Rev W. Lach-Szyrma drew up a comprehensive overview of the principal tenets of pluralism. His summary provides a useful and appropriate *point d'appui* for a larger set of conclusions regarding the long history of the subject.

He lived and wrote well within mainstream Western culture, then as ever the principal setting for pluralism. Also he wrote for a burgeoning pluralist audience: just over half the pluralist books published between the era of Galileo and the Great War appeared in the nineteenth century. The West developed the cosmologies which generated critical pluralism as well as the wealth and restlessness which sustained its development of theory and practice. *Aleriel* was written in English, historically the chief medium for pluralist literature, important contributions in many other languages notwithstanding.

The versatile Lach-Szyrma was well acquainted with astronomy, the natural science which has chiefly influenced pluralism; also he employed science fiction — the literary genre most close to pluralism. Scientists have often written pluralist fiction, for example Kepler, Sir Fred Hoyle, Carl Sagan; or they have been moved to action by it like Wernher von Braun, the pioneer rocket engineer, who somehow contrived to indulge his taste for the latest American pulps throughout a war in which he was a leading rocket engineer on the opposing side.

Lach-Szyrma's preface to *Aleriel* contains the following propositions:

> ★ *Pluralism is a normal speculation; Children ask: 'Are there people up there in those planets?'*

Whilst it may be a tall claim that all people have at all times speculated about pluralism, the Western intelligentsia has had a close interest in the subject, and taken sides on it, from the time of the pre-Socratic philosophers. Although this interest was intermittent until the Renaissance, it has subsequently remained on the intellectual agenda. Popular pluralism made its first dramatic appearance with the Great Moon Hoax of 1836, later with Lowell's Martian canals; it has been encouraged ever since by science fiction and popular science and is now commonplace.

An opinion poll, reported by the British *Sunday Telegraph* in late 1998 revealed that 28% of those interviewed believed 'there will be some signal of intelligent extraterrestrial life

A

CELESTIAL ATLAS,

DESIGNED PRINCIPALLY FOR THE USE OF

YOUNG LADIES,

CONTAINING 25 MAPS OF THE CONSTELLATIONS AND STARS VISIBLE TO THE NAKED EYE IN THE LATITUDE OF LONDON; ALSO A CATALOGUE OF THE STARS WITH THEIR RIGHT ASCENSIONS AND DECLINATIONS,

BY

WILLIAM FROST, F.R.A.S.

TEACHER OF

Writing, Arithmetic, Geography, Use of the Globes, & Mental Astronomy in Ladies' Schools,

3, CHATHAM PLACE, HACKNEY.

..e Author—"VIEW OF THE EARTH AND HEAVENS," 8s.——"PORTABLE TERRESTRIAL SPHERE," 1s.——"ARITHMETIC," 1s. 6d.——ARITHMETICAL AND ASTRONOMICAL TABLES, 6d.——"GEOGRAPHICAL AND BIOGRAPHICAL COPIES," 1s. 6d.

London:

PUBLISHED BY SIMPKIN & MARSHALL, AND YORK & CLARKE (LATE DARTON & HARVEY);

AND MAY BE HAD OF THE AUTHOR, 3, CHATHAM PLACE, HACKNEY.

64   *Ungendered pluralism: an early textbook for girls (c1840); astronomical and pluralist literature, particularly for young people, has generally had a good, gender-free, record.*

early in the next century'. This kind of enquiry, both pluralist and technical, would have had little meaning to most people throughout most of European history; now it makes sense and excites interest throughout the world. On the other hand, through most of human history the night sky, ultimate source of pluralist thoughts, was clear and familiar; by 2000 most urban dwellers live under the star-obscuring sodium-dome or its variants, often gaining their pluralist ideas from the media, or from fiction.

Whether or not children are natural pluralists is also debatable; if they are, then elemental pluralism is surely part of our mental structure. It is certain that more children are now exposed to pluralist assumptions and arguments from an early age than at any previous time. Even in the golden age of old pluralism, children's writers were generally circumspect about the subject (**64**).

Thomas Dick urged a pluralist line in his introduction to Elijah Burritt's textbook *Geography of the Heavens* (New York, 1845) with references to 'innumerable worlds which must exist throughout the immensity of space, the countless myriads of intelligences that people them...'; but most authors of books for young readers took a quieter line, even during the Lowellist era. H.H. Turner and Sir Robert Ball (**65**), professors of astronomy at Oxford and Cambridge respectively during Edwardian times, offered young readers

65   *An Irish Pluralist:*
Sir Robert Stawell
Ball, Lowndean
Professor of
Astronomy at
Cambridge, in the
University
Observatory c1895;
formerly Astronomer
Royal for Ireland,
Ball's popular
Story of the
Heavens *(1885)*
conveyed his
evolutionary
pluralism to a large
audience as did his
public lectures on
'Other Worlds'.

sensible advice. Turner suggested looking at both sides of the Martian canals question: 'I have not made up my mind myself'; he added that he was unconvinced by Wallace's anti-pluralism. Ball, a pluralist, went as far as saying Lowell's 'wonderful system of lines... would almost suggest that they had been laid down by intelligent guidance'.[1]

By 2000, however, responsible references to the possibility of aliens appear regularly in astronomical textbooks; there are even pluralist books for children such as *Is Anybody There?* (1998) by Heather Couper and Nigel Henbest. Science-based pluralism of this kind may sharpen the critical faculties necessary to cope with the flood of fictional pluralism produced for the young, particularly the pervasive space operas like *Star Trek* and *Star Wars*, complexes of books, magazines, computerised games and miscellaneous toys, apparel, and models. There can be few children in Europe and America, for example, who have not been exposed to the concept of a plurality of worlds. This alone constitutes one of pluralism's most remarkable historic advances.

66   *Tulse Hill observatory,
c1890: from this Heath
Robinson structure of
corrugated iron, onion
finials and a bent
stovepipe, Sir William
Huggins carried out his
pioneer work suggesting
'the brightest stars are like
our Sun, upholding and
energising centres of
systems of worlds adapted
to be the abode of living
beings'.*

★ *'Our Earth is singular in nothing'; Earth is not a special case: other planets bear
atmospheres, possess satellites,etc.*

The pluralist argument from analogy, and the principle of mediocrity have long been
mainstays of pluralist arguments; Fontenelle rested his case on 'the similarities of the
planets to the Earth which is inhabited'. Richard Bentley thought: 'every Fixt Star [is] of
the same Nature with our Sun'. Ebenezer Henderson's popular textbook *A Treatise on
Astronomy Displaying the Arithmetical Architecture of the Solar System* (1842) waxed technical in
the chapter 'On a plurality of worlds': 'the planets are in revolution round the Sun in one
direction', they had 'regular vicissitudes of the seasons' and were 'furnished with
atmospheres... [and] attended with moons', — and why? 'Planetary atmospheres must
certainly support planetary life... moons are indispensable to inhabited worlds at great
removes from the Sun...'

Simple analogy was, however, not enough in the post-Whewell era of scientific
pluralism which demanded and sought empirical proof for life on other worlds (**66**). At
first it seemed that Lowell had supplied this, but not so. A century later, scientists could
offer modest possibilities: ice on Europa; extra-solar planets, even possible systems of
planets. They employed analogy as advocated by J. S. Mill, as a source of suggestions to be
pursued, not as a species of rational proof.

67   *Direct democracy on Venus: four million Venusians gathered at an Athenian 'assembly in the Simerian mountains' listening to electrically-amplified debates; Lach-Szyrma's 'Letters from the Planets',* Cassell's Family Magazine, *1887.*

★ *It is unlikely that the Earth is 'the sole abode of vitality'*

That is, the argument from mathematical probability: 'surely in so vast a Universe there must be other worlds, other intelligences?' Sir John Herschel put the case for pluralist probability colourfully. Looking at the immensity of the Universe he asked: 'Would nature tip a hogshead to fill a wineglass?' Similar metaphors had, of course, been employed by Tsiolkovsky in the 1890s, and Metrodorus of Chios in the fifth century BC. Metrodorus was a disciple of Epicurus, whose atomism supplied an original basis for pluralist probability: the Universe contained an infinity of atoms, *ergo* an infinity of worlds.

The new pluralism, particularly in its most modern form, still employs probability, but rather more carefully, trying to quantify the likelihood of a plurality of worlds, for example by the Drake Equation or the 'guesstimates' of Bracewell, Asimov, Shklovskii, Sagan and many others.

There is another aspect of pluralist probability, the assertion that it is 'probable' that intelligent life exists on other worlds for one reason or another: as part of a divine or other cosmic plan; because of the panspermist scattering of life; or simply because it strikes a person as being likely, a kind of visceral pluralism.

In the old pluralism, probability arguments of this kind were generally vague and unquantified; even so great a mathematician as Leonhard Euler (1707-73) indulged in generalised sentiments: 'every planet, nay every one of the satellites has an equal right to

68   *Martian fauna: with 'radar domes' in the background, an illustration by Paul Hardy for Lach Szyrma's 'Letters from the Planets',* Cassell's Family Magazine, *1887.*

the same appellation [a world] it being highly probable that each... is inhabited' and again: 'we have an almost infinite number of worlds, similar to our Earth...'. More recently Professor Colin Pillinger FRS stated: 'It would be absolutely presumptious of us to believe we are alone in the Universe. Life on Earth is not just a one-off accident' — more cautious, but still probabilistic in essence.

To many pluralists probability arose from divine rationality: Nicholas of Cusa supposed 'that in every region there are inhabitants... all owing their origin to God.' Natural Theology put the case another way: 'a plurality of worlds is more worthy of the perfections of an Infinite Creator' (Thomas Dick, 1840). Of the celestial bodies, Ebenezer Henderson wrote: 'How can such mighty inanimate masses of matter adumbrate to the pleasure of the Creator? The moment it is admitted that [they] are peopled with intelligences capable of appreciating His wisdom, the Universe becomes liberated from a mass of incongruities.' The Catholic philosopher, Teilhard de Chardin (1881–1955) thought it 'unworthy of God... that the energies of matter' should be for the benefit of 'one single, living, human kind'.

In these ways the arguments from probability have often been a vehicle for another pluralist familiar: teleology, the assumption that worlds were created and inhabited for a purpose, perhaps divine, perhaps otherwise; the anthropic principle, for example, is coloured by this line of reasoning.

69　*Women of Ganymede,
one of the Jovian moons
— Stanley Wood's
illustration for
George Griffith's
'Stories of Other
Worlds',* Pearson's
Magazine, *1900:
anthropomorphic
pluralism in an Hellenic
mode.*

★ *'But, if there be life, what life?'* — *Life in the Universe may come in diverse forms.*

Pluralist history is rich with variations on the theme of biodiversity: Fontenelle's short, dark, and amorous Venusians; Whewell's 'boneless, watery, pulpy' Jovians; Swedenborg's spiritual planetarians; the moral hierarchies of Wright and Kant. In addition there are the multitudinous and imaginative creations of science fiction: Lach-Szyrma's own angelic Venusians (**67**); H.G.Wells' malevolent, cephalopodic Martians and Edgar Rice Burrough's beautiful, oviparous women of Mars. These are but a minute fraction of variations on this theme (**68, 69**).

A detailed taxonomy of imagined aliens would itself be mind-erodingly complex. Nevertheless, pluralist history offers some form to the apparent confusion. First, apart from the more wildly creative flights of imagination, there is a long tradition of attempting to adapt alien life forms to particular environments (**70**). Fontenelle and Whewell demonstrated this tendency; modern exobiology and fiction continues the task.

Secondly, although the majority of these life forms are projections of our own experiences, some genuinely novel forms have been suggested: sentient gas clouds

70 *Undesirable aliens: a double-headed monster in the 'semi-gaseous ocean' covering the planet 'In Saturn's Realm', part V of George Griffith's 'Stories of Other Worlds',* Pearson's Magazine, *1900.*

(Sir Fred Hoyle; *The Black Cloud*, 1957), stars containing 'plasmobes' — plasma based life[2]; and ultimately, sentient stars. A few philosophers and mystics have posited a spiritual Universe, with spirit inherent in matter — the theory of panpsychism advanced by G.T.Fechner and others. Some modern physicists have suggested that the Universe is essentially information, outwardly visible as matter. These theories recast the question of pluralism rather than address its central, traditional concerns.

Thirdly, the moral qualities — if any — of aliens were once considered closely, and to be of greater importance than their physical morphology or composition, although this element of pluralism has largely faded.

Finally, an assumption, possibly misplaced, of the essentially objective, rational nature of at least some aliens has given cartoonists and popular culture a useful ikon: the man or woman from Mars; the 'little green man' who asks innocent questions of our society, an ersatz answer to the problem of being able to see ourselves as others see us (**71**).

What might an alien make of arrangements ostensibly suffering rationality deficits: the division of the world into some two hundred sovereign states; income differentials; simultaneous public expenditure on hospitals and armaments? A depressing variation on this theme was uttered by the sagacious Dr Bernhard in the film *Berlin Express* (1948) 'Sometimes I think we will not get together on Earth until we find someone on Mars to hate.'[3]

71   The universal laws of economics? A possible explanation for the Great Silence; a Bill Hoechst pluralist cartoon from Parade *(1982). (Courtesy,* Parade*)*

"I'm afraid this will be our last trip here for a while . . . We've had massive budget cuts."

★ *'I rather suppose that Earth is at present merely an example of one phase of planetary development'; there may be a life-cycle on other worlds some of which bear primitive life whilst on others it may be far in advance of life on Earth, or perhaps have run its course and become extinct.*

These ideas, popular even before the theory of evolution by natural selection, are a refinement of the previous proposition. They were often employed by pluralists to support a range of assertions notably, during the era of old pluralism, about humanity's place in the great chain of being. To some pluralists humanity was at a staging post, usually of moral development. To Kant we were about half-way up the scale. Chalmers thought we might be the only fallen species. Whewell — who thought humanity unique — was willing to concede 'primitive life forms' on Mars. Baden Powell and Proctor, the latter popular at the time Lach-Szyrma wrote, advocated the evolutionary view that has long endured: species and cultures come and go, evolve and perish: that is the way of it (**72**). Eloquent fictional versions of this line were offered in *The Time Machine* and *The War of the Worlds* by the notable evolutionist, H.G.Wells.

★ *Lach-Szyrma 'hardly presumed to touch'... 'the theological question of God's dealings with the inhabitants of other worlds.'*

This significant side-stepping by a clergyman of the old and tendentious question of God's purposes in populating other worlds, and even more of the issue of the Incarnation which had coloured pluralism for centuries, suggests that although by the 1880s both questions were passing from the main pluralist agenda, sensitivities remained. Pluralism was by then adjusting

72   *More cultural detritus on the Moon: Stanley Wood's 1899 illustration for George Griffith's 'Stories of Other Worlds'; Pearson's Magazine, 1900.*

to the imperatives of science. Lach–Szyrma also wrote: 'I have... based my speculations on the known facts of astronomy.' He even invited readers to let him know if they found statements in *Aleriel* 'irreconcilable with the recent discoveries of inductive science'.

In the twentieth century pluralism continued to use information and ideas from 'inductive science'. If Karl Guthke and Steven J. Dick are correct, it has come in turn to direct science as a major axis of its discourse, offering it a focus, or mission. But still the ways and assumptions of the old pluralism continue to infuse the new. Analogy and probability remain important sources of pluralist argument. The reach of pluralists still tends to exceed the grasp of available science and technology. The many other subordinate logical and rhetorical ploys employed by pluralists over the years remain alive and well: inductive argument from particular to general; confusing necessary with sufficient conditions; inclining towards hypotheses with high explanatory power, but with a low possibility of empirical verification (UFOs and ancient-astronauts, for example); and many others in a treasure-house of examples for the students of texts on methodology and logic.[4]

Why has pluralism endured for so long? Is it wired in some way into our mentalities? Possibly, but it is more likely that it appeals because it gives us wonderful scope for analyzing ourselves, our beliefs and values in a uniquely *unearthly* setting. Optimists can talk of wise, immortal aliens who will teach us the arts of peace, clean technology, the conquest of illness. Pessimists warn us of malevolent, highly-evolved superbeings lacking emotions and moral restraint. A new literature is growing on the subject of what we might do, or ought to do, if contact is actually made with extraterrestrials; it tells us much about contemporary society and its singular ways. We are told variously that the discovery of extraterrestrial life will be the most dramatic finding of all time; that it will stagger us and change our self-image; or that people will soon adjust and become bored with the actuality of a plurality of worlds; evolution has taught us to be accommodating.[5]

Are we likely to discover extraterrestrial life in any case? In 1999 Sir Martin Rees, Astronomer Royal, put a middle-of-the-road case, drawing on the metaphysical assumptions of the old pluralism as well as the science of the new: 'I would give reasonable odds that there is life elsewhere, but much longer odds about actually being able to detect it.'

Meanwhile pluralism remains a fruitful source of dreams, of wonder, of distant goals — suggesting that it may turn out to be more pleasant to travel hopefully than to arrive. No understanding of this major, contemporary, cultural form is possible without a grasp of its historical context. History demonstrates that pluralism, at the cusp of two millennia, is the living outcome of continuities, cycles, repetitions and a long intellectual evolution.

Above all else, pluralism endures because it continues to connect us interestingly and effectively with the largest questions and profoundest mysteries: Are we alone? How should we behave towards other people, other species, other life-forms? Are we here for a purpose? What is the chief and highest end of man? And how are all things made for man? Answers or clues may, we think lie 'out there'.

We may usefully adapt the sentiments of Thomas Chalmers who was well acquainted with pluralism as well as with the big questions, and whose speculations remain appropriate:

> Who shall assign a limit to the discoveries of future ages? Perhaps the glass of some observer, in a distant age, may enable him to construct a map of another world, and to lay down the surface of it in all its minute and topical varieties. At present, for all their number, we have no knowledge of the moral system of any one of them. We are too distant to perceive the richness of their scenery and the bloom and luxuriance of their vegetation, to hear the hum of their mighty cities and the murmur of their populations; we have neither gazed upon their civil works nor savoured the fruits of their husbandry.

After two and a half thousand years of pluralism we can only say that perhaps one day we shall chance upon other worlds, other life forms, and other sentient beings and find out for ourselves; then again, perhaps not.

References
1   H. H. Turner: *A Voyage into Space*, 1915; R S Ball: *Star Land*, 1889.
2   See R. Shapiro and G. Feinberg: 'Possible Forms of Life in Different Environments', *Extraterrestrials*, eds Zuckerman and Hart, 1995.
3   The scriptwriter was Harold Medford; the actor Paul Lukacs.
4   See Michael Crowe, *op cit*, especially Chapter 11.
5   The results of possible contact with extraterrestrials was studied by some NASA-sponsored workshops on CASETI (Cultural Aspects of SETI) in 1990-91; see also Albert A. Harrison, *After Contact*, 1997.

Glossary

Alien (in pluralist usage) a being from another world, an extraterrestrial (qv).

Analogy similarity in details, properties, qualities.

Anthropocentrism perceiving humanity as the central purpose, point of reference, etc. in the Universe.

Anthropic principle set of theories (varying from 'weak' to 'strong' lines of argument) positing that the Universe is there because we are there to see it.

Antipluralism theories, doctrines, etc opposed to pluralism (qv).

Astro- see Exo-

CASETI Cultural Aspects of SETI.

CETI Communication (or Contact) with ETI.

Contact (in pluralist terms) an encounter between humans and ETs.

Copernican principle see Mediocrity.

Cosmicrobia that aspect of panspermism (qv) concerned essentially with basic life forms, eg microbes, travelling through space and seeding favourable locales.

Exo- prefix implying 'extraterrestrial' (qv); hence exobiology; exotheology; cf 'xeno-'. Occasionally the astro- prefix has been employed, eg 'astrobiology' the exobiology of other solar systems, or 'astrobotany' both usages owing much to the Soviet pioneer, G A Tikhov.

Extraterrestrial (adj) outside, or beyond Earth, (n) a being living beyond Earth.

ET an extraterrestrial being.

ETI extraterrestrial intelligence.

Mediocrity (principle of) — that Earth is a typical planet orbiting a typical star in a typical galaxy (implying that there may well be others). Also referred to as the Copernican principle.

Panspermism (or panspermia) a set of theories that life, most usually as microbes, bacteria, simple cells, etc., traverses, even suffuses the Universe (i) on rocks, meteors, comets — **lithopanspermism**; (ii) driven by radiation — **radiopanspermism**; (iii) on or in artificially constructed vehicles — **directed panspermism**.

Plenitude (principle of) — that which is possible will be (or will tend to be) realised.

Pluralism the theory that there is, has been, or might well be, life (intelligent or otherwise) elsewhere in the Universe; a sympathetic cast of mind towards this proposition.

Plurality of Worlds the doctrine that there are more worlds in the Universe than Earth alone.

SETI the Search for Extraterrestrial Intelligence.

Teleology doctrine that the Universe and all, or some entities within it, are created for a purpose, for an end; that the Universe displays evidence of purposeful design.

UFO Unidentified Flying Object (colloquially, 'flying saucer').

Ufology the study of UFOs — a field with its own argot, eg 'Befap' (Being from another planet); 'Manadim' (Man from another dimension); or 'Close encounters' — of up to four kinds — degrees and forms of contact with aliens.

Uniformity (principle of) - that the laws of Nature obtain throughout the Universe; that which brought about life on Earth could therefore do so elsewhere.

Xeno- prefix with the same purpose as 'Exo' (qv above); popular in the 1970s but tending to be replaced by Exo-. Hence Xenobiology, etc.

Xenology the 'science of [studying] aliens'.

Bibliography

Primary sources

The history of pluralism is to a considerable extent a chronicle of primary sources, many cited in the text of this book; some major milestones in the evolution of pluralism (a few titles abbreviated) are :

Giordano Bruno, *De L'infinito Universo et Mondi*, 1584
J. Kepler, *Somnium*, 1634
John Wilkins, *The Discovery of a World in the Moone*, 1638
Bernard le Bovier de Fontenelle, *Entretiens sur la Pluralité des Mondes*, 1686
Christiaan Huygens, *Cosmotheoros*, 1698
Thomas Chalmers, *A Series of Discourses on the Christian Revelation, Viewed in Connection with Modern Astronomy*, 1817
Thomas Dick, *Celestial Scenery*, 1837; *The Sidereal Heavens,* 1840
William Whewell, *Of the Plurality of Worlds, An Essay*, 1853
David Brewster, *More Worlds than One; The Creed of a Philosopher and the Hope of the Christian*, 1854
Rev Baden Powell, *Essays on the Spirit of Inductive Philosophy,* 1855
Camille Flammarion, *La Pluralité des Mondes Habités*, 1862; *Les Mondes Imaginaires et les Mondes Réels*, 1865
Richard A. Proctor, *Other Worlds than Ours*, 1870; *The Orbs Around Us*, 1872
William Miller, *The Heavenly Bodies: Their Nature and Habitability*, 1883
Joseph Pohle, *Die Sternewelten und ihre Bewohner*, 1884–85
Percival Lowell, *Mars*, 1895; *Mars and its Canals*, 1906; *Mars as an Abode of Life,* 1908
Alfred Russel Wallace, *Man's Place in the Universe*, 1903

Hybrid sources

From *c*1940; sources become specialist texts, partly of contemporary relevance, but partly historical sources because often overtaken in detail by the rapid developments in the field; they include:

H. Spencer Jones, *Life on Other Worlds*, 1940
Kenneth Heuer, *Men of Other Planets*, 1951
Walter Sullivan, *We Are Not Alone*, 1964, revised 1993
Carl Sagan & Iosif Shklovskii, *Intelligent Life in the Universe*, 1966
Ronald Bracewell, *The Galactic Club*, 1974
Ian Ridpath, *Worlds Beyond*, 1975; *Signs of Life*, 1975; *Life Off Earth*, 1983
John W. Macvey, *Alone in the Universe?* 1963; *Whispers From Space*, 1973
Duncan Lunan, *Man and the Stars*, 1974
James L. Christian, ed, *Extra Terrestrial Intelligence, The First Encounter,* 1976
Chris Boyce, *Extraterrestrial Encounter*, 1979
Donald Goldsmith and Tobias Owen, *The Search for Life in the Universe,* 1980
Isaac Asimov, *Extraterrestrial Civilisations,* 1980

Reinhard Breuer, *Contact with the Stars*, (English edition, 1982)
Joseph A. Angelo Jr, *The Extraterrestrial Encyclopaedia*, 1985
Joseph E. Baught, *On Civilized Stars*, 1985
John D. Barrow and Frank J Tipler, *The Anthropic Cosmological Principle*, 1986
Thomas R. McDonough, *The Search for Extraterrestrial Intelligence*, 1987

More recent texts, at or near state of the art include:

Ben Zuckerman and Michael H. Hart, eds, *Extraterrestrials, Where Are They?* (second
    edition, 1995);
Paul Davies, *Are We Alone?* 1995
Jean Heidmann, *Extraterrestrial Intelligence*, (English edition, 1995)
Seth Shostak, *Sharing the Universe*, 1998
Edward Ashpole, *Where Is Everybody? The Search for Extraterrestrial Intelligence*,1997
Barry Parker: Alien Life, *The Search for Extraterrestrials and Beyond,* 1998
Bruce Jakosky, *The Search for Life on Other Planets*, 1998
Heather Couper and Nigel Henbest, *Is Anybody There?* 1998, (for young readers)
Amir D. Aczel, *Probability 1,* 1998
Michael Kurland, *The Complete Idiot's Guide to Extraterrestrial Intelligence*, 1999

History of pluralism

Although one or two earlier works treated the history of pluralism this was usually by way of an introduction to a fuller, contemporary consideration of the subject. Recently histories of the subject *per se* have emerged as it has become embedded in modern culture. There are principally five; the first four in the list below constitute a 'trilogy plus one' systematically covering the subject from classical Greece to the late twentieth century:

Steven J. Dick, *Plurality of Worlds: The Origins of the Extraterrestrial Life Debate from
    Democritus to Kant,* 1982
Michael J. Crowe, *The Extraterrestrial Life Debate 1750-1900,*
Steven J. Dick, *The Biological Universe*, 1996
Steven J. Dick, *Life on Other Worlds,* 1998 (*The Biological Universe abridged and updated*)
Karl S. Guthke, *The Last Frontier*, English translation, 1990 of *Der Mythos der Neuzeit* (1983)
Arthur O. Lovejoy, *The Great Chain of Being: A Study of the History of an Idea*, 1936
I. F. Clarke, *The Pattern of Expectation, 1644-2001,* 1979
Chris Morgan, *The Shape of Futures Past*, 1980
Marjorie Hope Nicolson, *Voyages to the Moon*, 1948 and 1960
William G. Hoyt, *Lowell and Mars* (1976 and 1998)
B. Pullman, *The Atom in the History of Thought*, 1998

Science fiction

John Clute and Peter Nicholls (eds), *The Encyclopedia of Science Fiction*, 1993 — a fine work of scholarship which will offer the interested pluralist many leads
Brian Aldiss, *Trillion Year Spree,* 1973, 1986
Franz Rottensteiner, *The Science Fiction Book, An Illustrated History*, 1975
Philip Strick, *Science Fiction Movies*, 1976
David Kyle, *A Pictorial History of Science Fiction*, 1976
Phil Hardy, *The Encyclopedia of Science Fiction Movies,* 1984
John Clute, *Science Fiction, The Illustrated Encyclopedia*, 1995

Index